Field Geology
Illustrated

Terry S. Maley

Mineral Land Publications
P.O. Box 1186
Boise, Idaho 83701

(208) 343-9143

i

ii

PREFACE

Field Geology was designed to serve as a field reference to aid in recognizing, interpreting and describing geologic features at the outcrop. Emphasis is on the study of mesoscopic features that can be viewed in outcrop scale rather than large structures or landforms. This book is not an exhaustive or comprehensive treatise on the subject of field geology, but instead covers the information necessary to understand and describe most outcrops. Field procedures, rock and mineral descriptions and tables are omitted because they are covered in detail in several current texts.

Field Geology should be useful as a complementary text for any field-related geoscience course such as physical geology, field geology, petrology and structural geology. The detailed descriptions, illustrations and photographs of geologic features in their field setting will be particularly useful where field trips are not feasible.

This book is also intended for anyone who needs a good basic review of field geology including graduate students preparing for field mapping, professional geologists who wish to bring their skills up to date quickly and easily and even serious amateur geologists. Self study will be particularly rewarding because an interpretative sketch and detailed description is included with each photograph.

ABOUT THE AUTHOR

Terry Maley received B.S. and M.S. degrees in geology from Oregon State University and a Ph.D. in structural geology from the University of Idaho. Although he is well known for his widely successful mineral law books (3 books and 7 editions since 1977), he has been an avid observer, photographer and illustrator of geological features during 30 years of field work—amassing a collection of more than 10,000 color slides on the subject. He has authored more than 50 publications on geological features and structures and has also written the highly regarded *Exploring Idaho Geology.*

Before settling in the western U.S., he worked 8 years as a marine geologist participating in 15 world-wide expeditions with the Naval Oceanographic Office; he later worked for the Interior Department, the State of Idaho and several universities. Dr. Maley has served as president of the Idaho Association of Professional Geologists for many years.

CONTENTS

1 Stress and Strain

Stress and Strain Ellipsoids

Stress may be represented by three mutually perpendicular principal stress axes. **Stress ellipsoid** axes are proportional in length and parallel to the principal stress vectors. The stress ellipsoid corresponding to the standard state of hydrostatic stress is a sphere. This sphere, when subjected to a non-hydrostatic stress, would deform into a triaxial ellipsoid. This ellipsoid is the **strain ellipsoid** or the ellipsoid of deformation.

Pure Shear

The three axes of the strain ellipsoid correspond respectively in direction but not lengths to the three principal stresses (Fig. 1-1). In pure shear there is no rotation of the strained body.

Least principal stress axis = Pmin
Intermediate principal stress axis = Pint
Greatest principal stress axis = Pmax

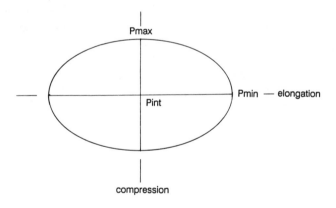

Fig. 1-1. Strain ellipsoid with pure shear (no rotation).

Simple Shear

In simple shear, the stress ellipsoid and the strain ellipsoid are non-coaxial. Simple shear deformation results when a torsional force or couple causes stress on a rock. Resolution of two couples has the effect of two principal stresses, Pmax and Pmin, acting at 45 degrees to the direction of shear (Fig. 1-2).

1

With progressive deformation, the strain ellipsoid will rotate about its intermediate axis (Pint) so that the direction of the maximum extension (Pmin) rotates towards the direction of shear and then the stress and strain ellipsoids will no longer be coaxial. The axis of rotation stays the same diameter as the original sphere. Structures formed from simple shear will have monoclinic symmetry. A set of sigmoidal extension gashes is a good example of a feature caused by simple shear.

Simple shear is perhaps the most important type of strain occurring in fault zones and near the contacts of intrusions. Any balanced system of stresses, which are caused by compressional, tensional or torsional forces, can be resolved into three principal stresses at right angles to each other.

Distinctions between Pure Shear and Simple Shear

Generally, it is difficult to distinguish pure shear from simple shear because the products look similar. In simple shear, the distribution of small feldspar grains at the rim of large feldspar porphyroclasts are asymmetrical because the small grains rotate away from the extensional principal stretching axis (Pmin). Tension gashes become asymmetric during simple shear because of rotation of the central first formed segment; if more than one gash, they tend to have en echelon arrangement. In pure shear, phenocrysts develop symmetrical tails and tension gashes open parallel to the least principal stress axis.

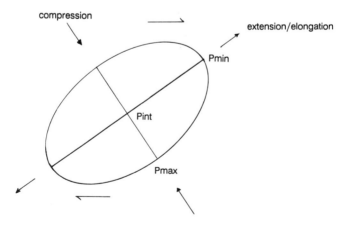

Fig. 1-2. Strain ellipsoid with simple shear (rotation).

2 Fractures

A **fracture** is a planar surface along which the rock has lost cohesion. The higher the crustal level the more brittle the rock and hence the more numerous the fractures. A **joint** is a type of extension fracture that has no displacement or movement parallel to the walls; whereas, a **fault** is a type of fracture that has experienced appreciable movement parallel to its walls. If no obvious evidence of faulting can be seen along a fracture such as offsets (Fig. 2-1), slickensides, gouge, brecciation, shearing, etc., then one should inspect the fracture surface for evidence that there was no displacement.

JOINTS

Joints may have steps on the fracture surface indicating that no displacement has occurred. **Plumose** texture, lack of grooves or **slickensides** also indicate whether a fracture is a fault or a joint. Because joints are generally formed by tension, there should be no evidence of compression along the fracture. A **fissure** is a fracture whose walls have moved apart. A **joint set** is a group of parallel fractures with a common origin. A **joint system** has two or more sets of joints in the same area that intersect at constant angles; if the sets intersect at approximately 90 degrees, they may be related as **conjugate** sets (Fig. 2-2).

Importance of Joints

Joints have a very significant effect on permeability, porosity and the structural integrity of large rock areas. They are the plumbing system for ground water flow and are the primary means of flow through metamorphic, igneous and sedimentary rock. Therefore they must be studied carefully prior to excavating or constructing projects such as dams or large structures. They also act, at least in part, as the underground plumbing system for ground water, hydrothermal solutions, mineralizing solutions, petroleum migration and accumulation. Studies of joints are also crucial for fluid or waste storage. Alteration, weathering, and solution of rocks is greatly facilitated by the access joints provide to the internal part of a rock body. Joints also control the access of igneous intrusions in the upper crust.

3

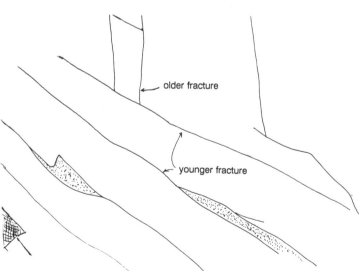

older fracture

younger fracture

Fig. 2-1. Two sets of fractures intersect in granite gneiss. The older set at the top of the photo is terminated by more recent movement along the lower set. Near Shoup, Idaho.

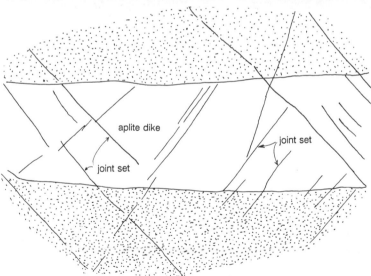

Fig. 2-2. Conjugate joints in aplite dike in granite. Because the dike is more competent than the granite, joints are more closely spaced and better developed in the dike. White Clouds, Idaho.

Joint Spacing or Density

Joint spacing depends on the rock type. In the same stress environment, more competent rocks will have greater joint density than less competent rocks. Where competent rock layers alternate with incompetent layers, joints may be confined to the competent layers (Figs. 2-3, 2-4, 2-5, 2-6 and 2-7). Joints also increase with proximity to the surface where rocks are under less pressure and are more brittle.

Origin of Joints

Most joint sets tend to cover broad areas despite changes in structure and rock types. The following categories are discussed in terms of their origin: (1) joints perpendicular to the earth's surface; (2) joints perpendicular to layering (similar to #1); (3) joints perpendicular to the cooling surface; (4) joints parallel to the topography; (5) joints perpendicular to linear structures; (6) diagonal joints; and (7) longitudinal joints.

Joints Perpendicular to the Earth's Surface

Joints perpendicular to the earth's surface are caused by compression from the weight of the overlying rock or by the vertical pressure caused by an upward moving pluton or both. Rocks tend to expand horizontally when subjected to vertical compression (Fig. 2-8).

Orthogonal sets of vertical joints are caused by horizontal extension along intermediate and least principal stress directions in response to compression along the greatest principal stress (Figs. 2-9 and 2-10). If only one joint set is developed, it would be the one perpendicular to the least principal stress direction.

Joints Perpendicular to Layering

Joints perpendicular to layering are genetically related to joints perpendicular to the earth's surface. These joints intersect layering such as bedding and cleavage (Figs. 2-11, 2-12 and 2-13) at high angles. Two vertical sets of joints at right angles to one another (orthogonal joints) are common. In flat-lying layered rock, these perpendicular joints are indistinguishable from those found perpendicular to the earth's surface mentioned above.

Joints Perpendicular to the Cooling Surface

When a homogenous rock is cooled, it will contract and tension joints will form. In a lava flow, **columnar joints** perpendicular to the cooling surfaces will develop (Figs. 2-14

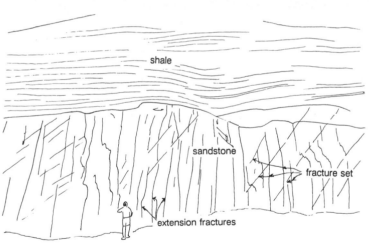

Fig. 2-3. Shale overlying sandstone in Colorado. Note the two sets of fractures developed in the more competent sandstone but not the shale.

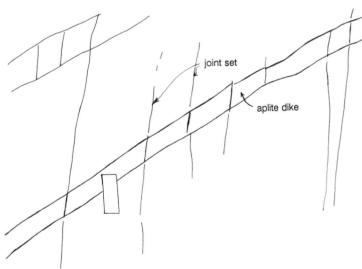

Fig. 2-4. Joint set in granitic rocks intersects aplite dike. The dike is more competent than the granitic host rock so that the joints are more closely spaced and better developed in the dike. White Clouds, Idaho.

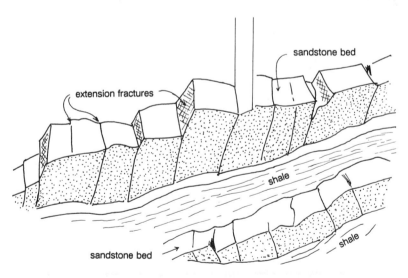

Fig. 2-5. Interbedded Cretaceous sandstones and shales with extension fractures in the more competent sandstone beds. West of Bend, Oregon.

9

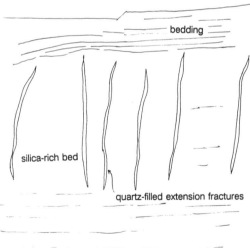

bedding

silica-rich bed

quartz-filled extension fractures

Fig. 2-6. Quartz-filled extension fractures in quartzite preferentially formed in the more competent or silica-rich bed. Near Stanley, Idaho.

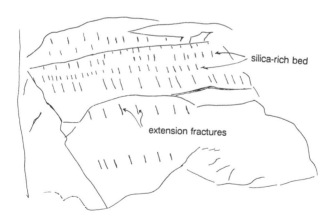

silica-rich bed

extension fractures

Fig. 2-7. Paleozoic quartzite with extension fractures developed preferentially along the more competent or silica-rich beds. Near Challis, Idaho.

Fig. 2-8. Well developed set of vertical fractures, enhanced by spheroidal weathering, result in deep weathering of fractured areas and rounding of forms. Silent City of Rocks, Idaho.

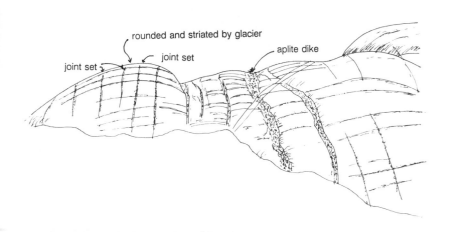

rounded and striated by glacier

joint set

joint set

aplite dike

Fig. 2-9. Close-up of granitic rock ground down by passing glacier. Note the vertical joint set that is intruded by aplitic dikes. Wallowa Mountains, Oregon.

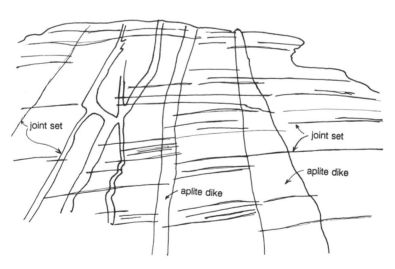

Fig. 2-10. Glaciated outcrops of granitic rock with orthogonal vertical joints. Wallowa Mountains, Oregon.

14

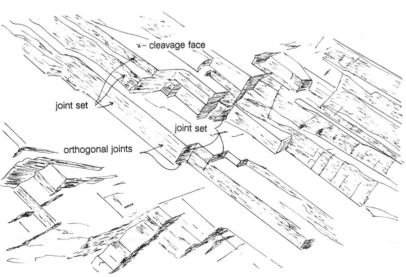

Fig. 2-11. Orthogonal joint sets are perpendicular to layering (cleavage) in quartzite. Ellis, Idaho.

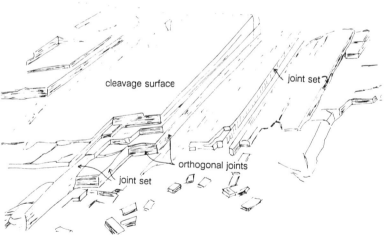

Fig. 2-12. Orthogonal joint sets are perpendicular to the foliation of the Precambrian quartzite. Middle Mountain, Idaho.

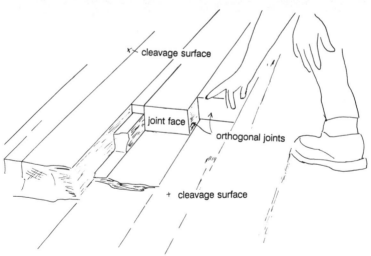

cleavage surface

joint face

orthogonal joints

+ cleavage surface

Fig. 2-13. Orthogonal joint sets in micaceous quartzites perpendicular to the foliation surface. At this quarry, near Middle Mountain, Idaho, brick-shaped blocks of quartzite can easily be extracted.

17

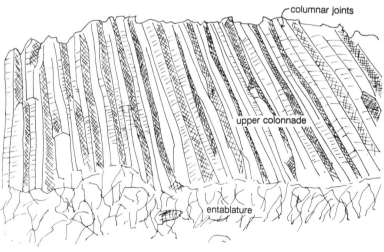

Fig. 2-14. Columnar joints in the upper colonnade of Columbia River Basalt flow in eastern Oregon. Near Picture Gorge.

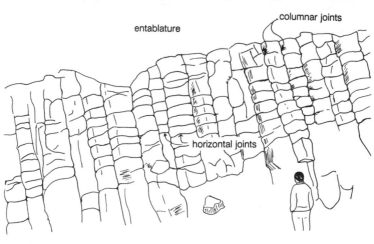

Fig. 2-15. Columnar joints in lower colonnade of Columbia River Basalt. Note the horizontal joints, independent with each column, formed after the development of the columnar joints. South of Coeur d'Alene, Idaho.

and 2-15). In a cooling intrusive body (pluton), tension fractures will also form in response to contraction. These fractures will form perpendicular to the intrusive contact which is also the cooling surface.

Columnar joints are common in basalt, intermediate volcanics and welded tuffs. Rapid cooling causes polygonal columns to form at both the top and base of the lava flow so that two **colonnades** will develop. A colonnade is the zone, either at the top or bottom of a flow, that is characterized by columnar jointing. Two colonnades may be separated by an **entablature** zone.

Joints Parallel to Topography

Erosional unloading causes extension joints to form parallel to the topography (Fig. 2-16). Rock expands vertically when the weight of the overlying rock is removed (Fig. 2-17). This process is referred to as **sheeting** or **exfoliation**. It tends to be best developed in granitic rock such as that exposed at Yosemite National Park (Fig. 2-18), but also occurs in quartzites and sandstones. The rocks are fractured into successive layers about 0.5 to 2 m thick. The layers are successively thicker with depth and generally do not exist far below the surface.

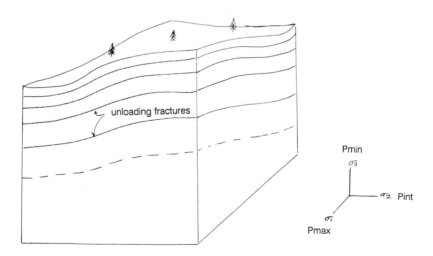

Fig. 2-16. Relationship of principal stress directions relative to unloading fractures.

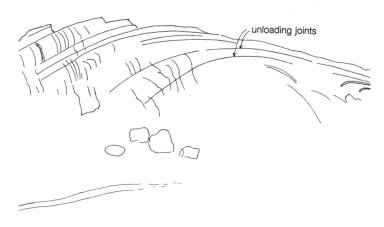

Fig. 2-17. Unloading joints in granite appear to thicken away from the dome. Silent City of Rocks, Idaho.

21

Fig. 2-18. Unloading joints, parallel to the topography, in the Sierra Nevada Batholith, northern California.

Joints Perpendicular to Linear Structures

Cross joints are perpendicular to linear structures such as fold axes and primary flow lineation in plutons (Fig. 2-19). They are also called **ac joints** because they tend to lie in the ac fabric plane. In plutonic rocks the flow lines represent the direction of maximum extension in the crystallizing melt. As a late-stage event in this flow and elongation, cross joints form perpendicular to the lineation and are the first fractures to develop in the cooling pluton. These joints are commonly filled with pegmatite and aplite dikes and hydrothermal quartz.

Diagonal Joints

Diagonal or **transverse joints** transect the regional structure or flow lines of plutons at a 45 degree angle. These joints in plutons may be occupied by aplitic dikes as well as veins of hydrothermal minerals such as chlorite, fluorite and epidote.

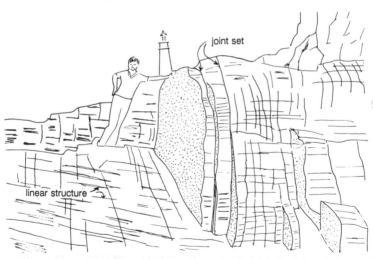

Fig. 2-19. Set of cross joints perpendicular to both strong secondary lineation and foliation. Near Portland, Maine.

Longitudinal Joints

Longitudinal joints strike parallel to the trend of flow lines or the regional structural trend. These joints may be occupied by aplite dikes, pegmatite dikes, basalt dikes and quartz veins.

Hydraulic Joints and Fracture Propagation

Fluid plays an important role and may generate joints by **hydraulic fracturing**. This type of fracture is most likely to occur in the upper crust where there is much water and the crust is brittle. However, such joints also tend to occur at depths greater than 5 km where the fluids cannot readily escape. Magma may also function as a fluid and cause hydraulic fractures. These fractures commonly attenuate or change direction at the contact between two rock types.

Joint Patterns

Radial joints are arranged in a radial pattern around a volcanic center such as volcanic vents underlain by large magma chambers. Withdrawal of magma from the chamber may cause collapse of crust over the chamber. **Ring fractures** may form a cylindrical pattern around a volcanic center. Radial joints and ring fractures are also commonly found together around a volcanic center.

Joint Surfaces

Joints commonly originate as a set of subparallel **en echelon fractures** (Fig. 2-20). With continued tension, a single continuous fracture may form by breaking across the connecting slabs between the en echelon fractures. Such a joint surface will be rough and have a stepped surface. If steps are found on the fracture face, you have reliable evidence that little or no displacement has occurred.

Plumose Structures

Joints in cohesive and fine grained rocks such as argillites tend to have **plumose** structures on the joint face (Figs. 2-21 and 2-22). The joint starts at the center of the plume and propagates along the plume lines to the fringe on both sides. The plume lines may terminate at small en echelon fractures at the fringe.

Conjugate Sets or Conjugate System

A **conjugate set** is a system of joints consisting of two or more sets which appear to have formed simultaneously (Fig. 2-2). They may form parallel to the shear planes of the strain

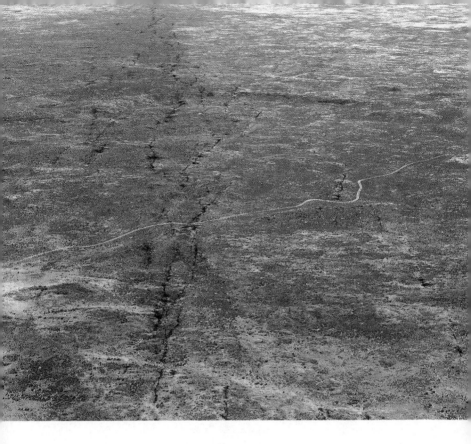

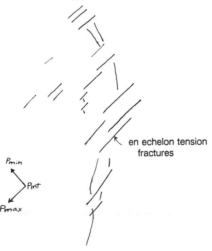

Pmin

Pint

Pmax

en echelon tension
fractures

Fig. 2-20. En echelon fractures along the Great Rift very likely join at depth to form a continuous fracture. South-central Idaho.

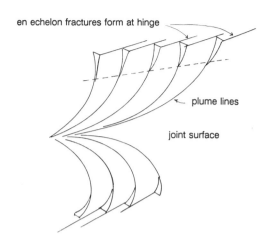

en echelon fractures form at hinge

plume lines

joint surface

Fig. 2-21. Plume structure on joint surface.

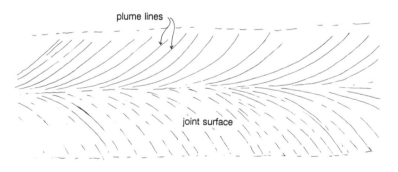

plume lines

joint surface

Fig. 2-22. Plumose structure on joint surface in sandstone. North of Boise, Idaho.

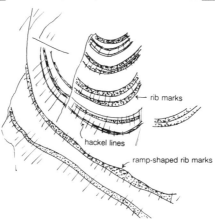

Fig. 2-22A. Excellent example of curvilinear features called **rib marks** or **arrest lines** on a joint face. These ramp-shaped rib marks cross the lines of hackle or plume lines at a right angle.

28

Fig. 2-23. Set of quartz-filled joints in Precambrian quartzites. Note the ripple marks preserved on the bedding surface in the lower part of the photograph. The joint set displays separation at each joint. Joints have an average spacing of 1 m. Near Ellis, Idaho.

ellipsoid and intersect at an acute angle. In some cases each en echelon fracture has a sigmoidal (S-shaped) profile.

Joint Coatings and Fillings

Joints may be coated with a variety of minerals such as iron oxides, manganese oxides, epidote and carbonates. They may also be filled with dikes or vein minerals. If the joints contain vein minerals, they probably originated as open fractures. Veins may be filled with quartz (Fig. 2-23), calcite or some ore mineral. Joints filled by dikes may have dilation features. Comb structures, vugs and matching walls are evidence of open fractures.

Joints in Plutons

Joints in plutons are strongly influenced by pluton boundaries (Figs. 2-24, 2-25, 2-26 and 2-27). Early fractures form during final crystallization and are related to flow banding.

Fractures are more common and closely spaced at the margins of plutons. Sheet or unloading joints form subparallel to the surface topography and the spacing becomes closer towards the surface. Orthogonal sets of shrinkage or tension joints form perpendicular to the cooling outer surface of the pluton. These joints also tend to have closer spacing towards the surface. Joints in plutons are commonly filled with hydrothermal minerals, pegmatite, aplite and quartz.

FAULTS

A fault is a fracture with appreciable displacement parallel to the fracture plane or fault surface. The **displacement,** also referred to as **offset** or **slip**, is the distance on a fault surface between two originally adjacent points on opposite sides of a fault. The **fault surface** is the surface along which fracture and displacement have occurred. A **fault zone** is typically tabular in shape (Fig. 2-28), along which shearing, brecciation, crushing and foliation may be observed. This zone may also be

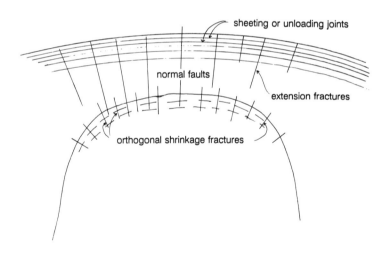

Fig. 2-24. Cross section of pluton intruded into the crust. A typical distribution and relationship of joints to the pluton are shown. Unloading joints are formed roughly parallel to and near the surface. Extension fractures and normal faults are formed in the crust above the pluton; they are caused by the vertical pressure exerted on the crust by the upward-moving pluton. Shrinkage or cooling fractures form in the upper margin of the pluton.

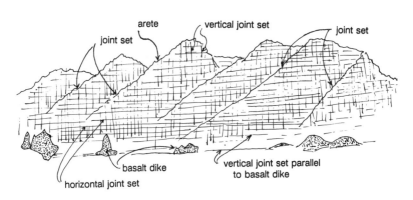

Fig. 2-25. Joints in Wallowa Mountains, Oregon having major effect on the glaciated granitic rocks. The sawtooth appearance of the ridge is accentuated by two sets of vertical joints. Remnants of a basalt dike, which apparently filled a joint parallel to the plane of the photograph, can be seen at the base of the mountain.

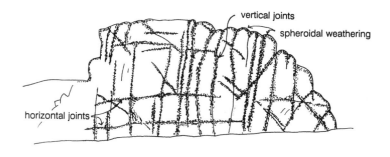

vertical joints

spheroidal weathering

horizontal joints

Fig. 2-26. This granitic outcrop is cut by three prominent joint sets: two orthogonal vertical sets and one horizontal set of unloading joints. The two vertical sets were developed first while the maximum principal stress direction was vertical. Much later when the erosion of overlying rock brought the rocks shown in the photo to the surface, the vertical stress direction became the minimum principle stress. Then the horizontal expansion joints formed. Silent City of Rocks, Idaho.

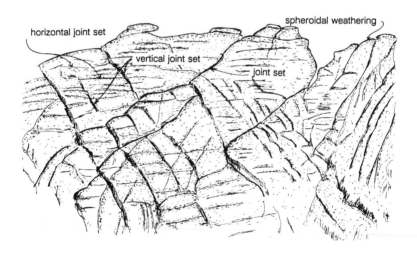

Fig. 2-27. Horizontal and vertical joint sets in granitic rocks. Note how surface weathering has enhanced the joints. Silent City of Rocks, Idaho.

referred to as a shear zone (slabby) (Fig. 2-29), a breccia zone (fragmental) (Figs. 2-30 and 2-31), or crush zone (gouge) (Fig. 2-32), depending on the dominant character. A fault zone may range from less than a centimeter to several hundreds of meters in width. The **hanging wall** on a fault is the upper wall of an inclined fault, or the wall on which the miner hangs his lamp. The **footwall** is the lower wall of an inclined fault; it is the wall the miner stands on.

Brittle Shear Zones

In the upper 10 km of the earth's surface, faults are caused by brittle fracture and displacement. Fault zones are characterized by a variety of brittle deformation features such as slickensides (Figs. 33A and 33B), brecciation, shearing, crushing, gouge and cataclastic rocks (where the individual grains have lost cohesion). **Cataclasite** is a structureless, nonfoliated rock with very few fragments visible to the naked eye. A **cataclastic rock** is a preexisting rock which has been broken to angular fragments by mechanical processes. Cataclasites tend to be structureless rock powder. **Cataclastic flow** is the flow of fine-grained material produced by brittle deformation rather than ductile flow that produces mylonites.

Pseudotachylyte is a glassy or cryptocrystalline material that looks like dark volcanic glass and forms by melting along brittle fault planes from heat generated by frictional sliding (Fig. 2-34). Pseudotachylyte generally occurs in low porosity

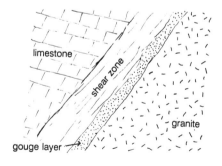

Fig. 2-28. Cross section of a shear zone that brought limestone into contact with granite. Note that shearing of granite at high crustal levels (where rocks are brittle) produces gouge.

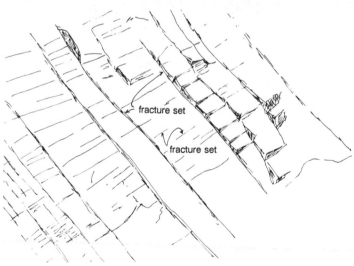

fracture set

fracture set

Fig. 2-29. A set of parallel faults in granite, each with minor displacement but cumulatively large displacement. Near Lowman, Idaho.

35

Fig. 2-30. Rounded pebble from the Big Wood River. Limestone breccia filled with white carbonate cement. Near Shoshone, Idaho.

Fig. 2-31. Brittle fault in quartzite. For up to 3 m on either side of the main fault, fractures cut the quartzite approximately parallel to the fault and spaced several centimeters apart. Middle Mountain, Idaho.

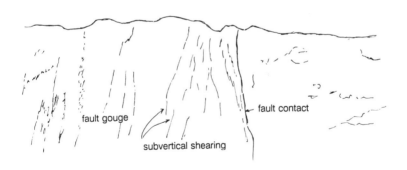

fault gouge

subvertical shearing

fault contact

Fig. 2-32. Vertical fault in andesite with thick gouge zone.

Fig. 2-33B. Fault striations on the lower block of a thrust fault. Northeastern Nevada.

Fig. 2-33A. Slickensides caused by offset along a steeply dipping vertical fault plane. The black pen is approximately parallel to the fault striations and the movement direction of the fault. Near Challis, Idaho.

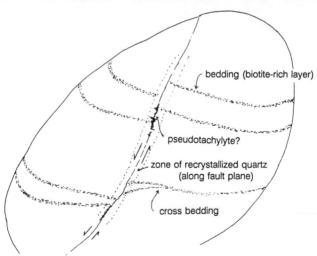

bedding (biotite-rich layer)

pseudotachylyte?

zone of recrystallized quartz
(along fault plane)

cross bedding

Fig. 2-34. Well rounded quartzite pebble from Big Wood River near Shoshone, Idaho. Laminations defined by biotite display cross bedding which is offset by a fault.

rocks such as gabbro, gneiss and amphibolite rather than in porous sedimentary rock. Contacts defining the outer boundaries of the pseudotachylyte are thin and sharp. Pseudotachylyte is typically injected into small oblique fractures cutting the wall rock.

Ductile Shear Zones

At crustal depths greater than 10 km, faulting is accomplished by ductile deformation. These faults are **mylonitic** or schistose, but not cataclastic because there is no loss of cohesion among the grains. As a general rule, faulting in the upper 5 km of the earth's crust is characterized by gouge and breccia. At crustal depth of 10 to 15 km deformation products tend to be crush breccia and cataclasites. At depths greater than 15 km the rocks may be foliated and mylonitic.

Mylonites are strongly foliated metamorphic rocks with high ductile strain and only partial recrystallization (Figs. 2-36 and 2-37). They form only as a result of ductile deformation and have the fabric and texture of extreme ductile deformation. Compared to adjacent undeformed rocks, mylonite has a small grain size and stronger foliation. **Protomylonite** is partially mylonitized rock.

Porphyroclasts are large crystals that have survived the reduction in grain size due to shearing and tend to have asymmetric tails in mylonites. The small grains in the matrix may have granoblastic texture with interlocking grain boundaries, forming 120 degree triple junctions or very sutured boundaries. Augen gneiss and mylonitic gneiss develop asymmetric simple shear tails on porphyroclasts with most of the material in the tails derived from the porphyroclasts. **Porphyroblasts** are large grains that grew during deformation and metamorphism; whereas, **porphyroclasts** are large relict grains with little or no recrystallization.

Evidence of Displacement. Strong deformation in ductile shear zones forms simple strongly foliated fabric by realigning and rotating existing structures. Such linear structures as rotated boudins, drag folds (Figs. 2-37 and 2-38), rootless folds, sheath folds and isoclinal folds are formed. The axes of these folds would be perpendicular to the last movement direction. However, stretching lineations are normally parallel to the movement direction. Structures, large and small, may be displaced. Rotated folds and boudins will give a vergence direction to show relative displacement.

40

Fig. 2-35. Strongly foliated mylonitic diorite in ductile fault zone. This is a typical exposure along the western Idaho suture zone where the foliation dips steeply to the east. The fault zone is 1 to 2 km wide. North of McCall, Idaho.

Fig. 2-36. Close-up of strongly foliated mylonitic diorite in ductile fault zone. Near Orofino, Idaho.

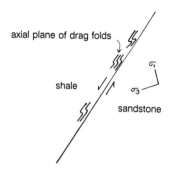

Fig. 2-37. Drag folds occur in the incompetent shale and indicate relative displacement along the fault.

Ductile shear zones, characterized by strongly foliated gneissic rocks, separate mesozonal and deep plutons from the country rock. Drag folds and rotated boudins indicate the relative displacement between the pluton and the country rock.

Shear zones with sharp boundaries crossing older shear zones can look like cross bedding particularly in mylonites (Fig. 2-39). Structures may also look like graded bedding. Ductile shear zones may narrow when passing into brittle rocks such as granite.

TYPES OF FAULTS

Dip-Slip Faults

In **Normal faults** the hanging wall block moves down relative to the footwall block (Figs. 2-40 and 2-41). This fault tends to fracture along a plane that dips 55 to 70 degrees, with 60 degrees the most common. Because a normal fault forms by horizontal extension, it may have porous breccia within the fault zone. Porous zones provide the plumbing systems for hydrothermal solutions that may precipitate minerals and form veins. The dip may shallow in shale or other subhorizontal bedding and become a detachment fault. These faults may also flatten in the zone of ductile flow.

Thrust or Reverse Faults

In **thrust or reverse faults** the hanging wall block moves up relative to the footwall block (Figs. 2-42 and 2-43). Thrust faults generally dip less than 45 degrees and are characterized by low dips, large displacements and older rocks over younger.

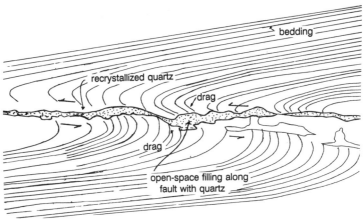

Fig. 2-38. Quartz-filled fault in laminated quartzite. Note the drag folds on both sides of the fault indicate the relative displacement. Middle Mountain, Idaho.

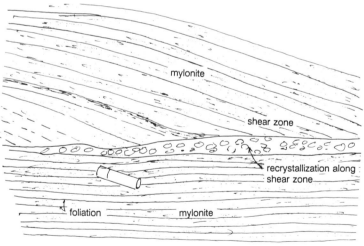

Fig. 2-39. Shearing in the mylonitic diorite gives the appearance of false cross bedding. Near Orofino, Idaho.

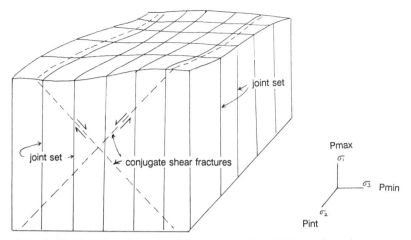

Fig. 2-40. Relationship of orthogonal vertical joints and conjugate shear fractures to the principal stress axes.

Fig. 2-41. The 1983 Borah Peak earthquake registered 7.3 on the Richter scale and resulted in a 35-km-long fault scarp. Several subvertical normal faults with a cumulative vertical displacement of 3 m disrupted the gravel road. Northwest of Mackay, Idaho.

45

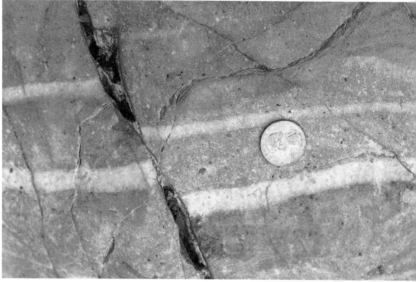

Both photographs on this page were taken about 10 m apart in the same marble unit. The top photograph shows a normal fault with ductile deformation. Notice the drag on the lenticular layers that indicate the relative movement along the fault. The bottom photograph shows a normal fault with brittle deformation. Note the cavities along the rupture plane that may have originated from dissolution and (or) separation along the fault. These two faults apparently occurred when this marble unit occupied two different crustal levels. The ductile deformation occurred during deep burial and the brittle deformation occurred much later and at a much higher crustal level. Marble Canyon in Death Valley, California.

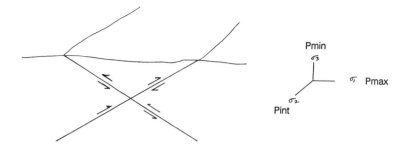

Fig. 2-42. Relationship of conjugate thrust faults to the principal stress axes.

They may be associated with evidence of horizontal compression such as folds with steeply inclined axial planes and hinges parallel to the strike surface. These faults may also be deformed by folds that trend parallel to the strike of the fault. The vergence direction of the folds should be in the direction of overthrusting or overfolding.

Overthrust Faults

An **overthrust fault** is a thrust with a large displacement where the upper block has moved more than 5 km. The unit above the fault plane is commonly referred to as a thrust sheet, nappe, allochthone or upper plate.

Low angle faults may be parallel to bedding and difficult to recognize. Evidence of high fluid pressure includes clastic dikes cutting into the hanging wall, hydrofracturing and missing strata.

Detachment or decollement faults are characterized by a bedding plane zone of slip or a zone of discontinuity that is overlain by folds and faults. Decollement faults tend to occur in ductile units where rocks of low permeability have underlying rocks saturated with fluids. Evidence for such faults includes: (1) missing strata; (2) faulted and folded overlying rocks; (3) undeformed underlying rocks; (4) cleavage and lineations parallel to the fault zone; and (5) an erosional fault with conglomerate and sedimentary breccia overridden by the thrust sheet.

Strike Slip Faults

Right lateral or **dextral faults.** When looking across the fault, the block on the other side of the fault moved horizontally to the right.

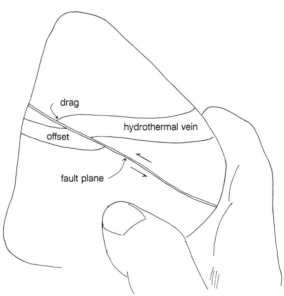

Fig. 2-43. Gray quartzite pebble from the Big Wood River has pure-white hydrothermal quartz vein offset by reverse fault. Near Shoshone, Idaho.

Left lateral or **sinistral faults**. When looking across the fault, the block on the other side of the fault moved horizontally to the left.

En Echelon Fractures

En Echelon tension fractures (gashes) may indicate the direction of displacement by two features (Figs. 2-44 and 2-45): (1) sigmoidal profiles of fractures, and (2) offset direction of vein-filled fractures.

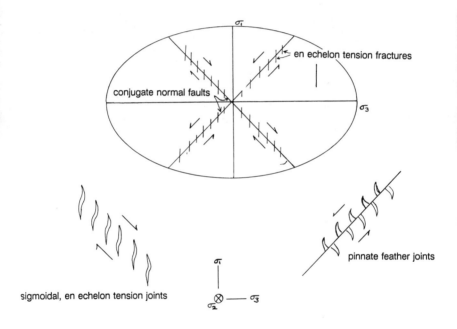

Fig. 2-44. The relationships of conjugate normal faults and en echelon tension fractures to the strain ellipsoid and the principal stress axes. Also shown is an enlarged view of sigmoidal, en echelon joints and pinnate or feather joints.

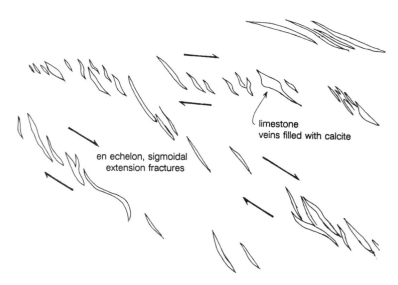

en echelon, sigmoidal
extension fractures

limestone
veins filled with calcite

Fig. 2-45. Several sets of sigmoidal, quartz-filled, en echelon fractures in quartzite. West of Ketchum, Idaho.

50

Relationship of Common Structures Relative to the Principal Stress Axes

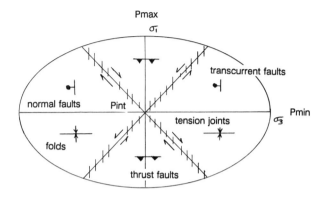

Fig. 2-46. The relationships of transcurrent faults, normal faults, thrust faults, tension joints and fold axes to the strain ellipsoid and the three principal stress axes.

Slumps

A **slump** is a nontectonic fracture and is not genetically related to the fractures discussed above. Because a slump is a form of mass wasting rather than a tectonic fault, one should be aware of the distinctions in the field.

A slump is a mass of rock, generally unconsolidated or partially unconsolidated, that slides downhill under the force of gravity (Fig. 2-47). The associated shear plane is normally concave upward like a spoon so that the slump block rotates backward on the plane of failure.

Typically the slump block is broken into a set of stepped slumps, each of which is rotated along a shear plane that is tangential to the lowest or main shear plane. Among the distinctive features associated with slumps are: (1) main scarp (Fig. 2-47);(2) tension cracks in the vicinity of the main scarp; (3) ponded water; (4) rotated blocks and features (Fig. 2-49); and (5) hummocky toe (Fig. 2-50).

Important causes of slumps include the following: (1) excess pore water pressure; (2) load added to slope; (3) undercutting at the base of slope; and (4) clay layers along potential slip plane.

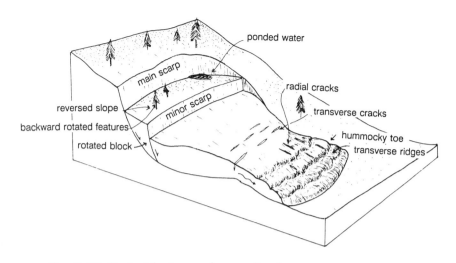

Fig. 2-47. Typical features of a gravity slump.

Fig. 2-48. Headwall scarp and series of slump blocks on Bliss landslide. Bliss, Idaho.

Fig. 2-49. Rotated slump block within the Bliss landslide. Bliss, Idaho.

Fig. 2-50. Earthflow at lower end of the Bliss landslide temporarily covered the entire channel of the Snake River. A temporary lake formed upstream which raised the water level enough to carve a new channel to the south of the original channel. Looking north at road removed by new channel. Bliss, Idaho.

3 Rock Cleavage

Definitions of Rock Fabric and Related Terms

Rock Fabric is the internal arrangement of the elements or grains of which a rock is composed; it includes texture and preferred orientation as well as the relative size, shape, homogeneity and arrangement of the constituent crystals.

Penetrative fabric is developed throughout a rock; whereas, **non-penetrative** fabric is confined to a discrete layer or plane such as a fault surface.

Planar fabric is expressed as a series of parallel planes of cleavage or foliation.

Primary fabric reflects the origin of a rock either as sedimentary or igneous.

Secondary fabric reflects deformation of the primary fabric and creation of new fabric elements.

Foliation includes all planar structures, both primary and secondary, excluding fractures, that are marked by parallel fabric elements. Examples of foliation include (1) platy minerals and lenticular mineral aggregates; (2) sedimentary bedding; (3) cleavage or schistosity; (4) compositional layering; and (5) textural layering.

Slate is a compact, fine-grained, metamorphic rock derived from rocks such as shale and volcanic ash; it has the property of fissility along parallel planes independent of the original bedding.

Phyllite is intermediate between slate and schist.

Schist is a medium- to coarse-grained metamorphic rock with parallel or subparallel orientation of the micaceous minerals which predominate over nonplaty minerals.

Gneiss is a coarse-grained rock in which bands or layers of granular minerals alternate with bands or layers of schistose minerals.

Cleavage

Cleavage is a type of secondary foliation occurring in recrystallized metamorphic rocks; it is expressed as a tendency of rocks to part along parallel planes. Cleavage consists of closely spaced parallel planes of weakness or potential parting. There is no rupture along cleavage planes; however, no matter how closely spaced, joints are definite surfaces of rupture.

There are two types of cleavage: (1)**continuous cleavage** where all platy minerals are parallel throughout the rock; and

(2) **spaced cleavage** where a set of cleavage surfaces are evenly spaced, with unaffected intervening spaces.

Continuous Cleavage

Continuous cleavage includes **slaty cleavage** and **schistosity**. This cleavage is developed by new minerals growing in response to compression where new minerals are aligned and elongated so that the planes of cleavage have a common orientation. Continuous cleavage tends to develop more easily in shales and tuffs than sandstone or basalt.

Slaty cleavage is caused by the parallel arrangement of platy minerals in slates. This parallel foliation of fine-grained, platy minerals (chlorite and sericite) is formed in a direction perpendicular to the direction of compression. Slaty cleavage has very smooth and closely spaced cleavage planes ranging from less than 1 mm to 2 mm. Slaty cleavage tends to form in fine-grained rocks such as shales (Figs. 3-1 and 3-2), mudstones and tuffs.

Schistosity generally refers to coarsely foliated continuous cleavage as opposed to the finer minerals in slaty cleavage. Schistosity is common in high-grade metamorphic rocks. In the slaty cleavage of slates, rock splits along smooth flat slabs; however, in more coarsely crystalline rocks such as schists, rock tends to split along undulating or corrugated surfaces.

Spaced Cleavage

Spaced Cleavage is uncleaved rock separated by cleavage planes (Fig. 3-3). **Microlithons** are uncleaved layers between cleavage surfaces. Spaced cleavage includes crenulation cleavage, pressure solution cleavage and fracture cleavage.

Crenulation cleavage or **strain-slip cleavage** is a type of spaced cleavage along which an earlier cleavage or schistosity (S1) has been deformed by the development of a new cleavage (S2) (Fig. 3-4). In other words, the micas that make up the earlier continuous cleavage (S1) are partly rotated to parallelism with the new spaced cleavage (S2); this generally occurs in the hinge areas of folds. New mica growth may also occur along the new cleavage (S2). Crenulation cleavage typically overprints an existing cleavage by crenulating the earlier cleavage.

Crenulation cleavage may also be referred to as strain-slip cleavage or slip cleavage where the new spaced cleavage (S2) is sufficiently developed to cause differential movement along a nearly parallel series of closely spaced shear or fault planes

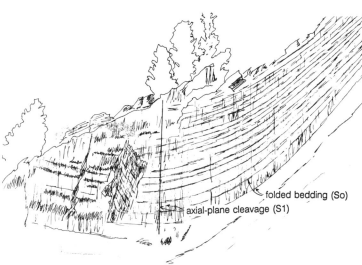

folded bedding (So)

axial-plane cleavage (S1)

Fig. 3-1. Synclinal fold in shale with axial-plane cleavage. British Columbia.

Fig. 3-2. Close-up of limb of fold shown in Fig. 3-1. Notice how the cleavage intersects the bedding planes of the shale.

that form the new spaced cleavage (S2). Between each pair of spaced cleavage or shear planes (S2), the original crenulation cleavage may be deformed into sigmoidal folds.

Pressure-solution cleavage produces spaced cleavage by dissolving the most soluble components of the rock and leaving irregular thin layers of residual minerals (Fig. 3-5). Therefore a series of parallel planar surfaces are formed by the dissolution and removal of material such as quartz, feldspar and carbonate. Consequently, the cleavage bands have a different composition than the intervening layers.

Fracture cleavage is a somewhat questionable type of spaced cleavage. The term is generally applied to a type of axial-plane cleavage that occurs in silica-rich rock such as sandstones under low-grade metamorphic conditions.

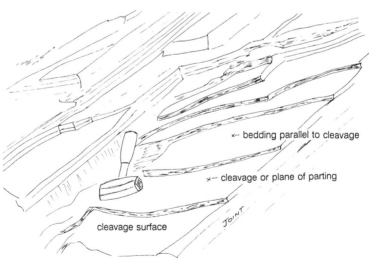

x— bedding parallel to cleavage

x— cleavage or plane of parting

cleavage surface

JOINT

Fig. 3-3. Quarry where quartzite is easily split along spaced cleavage. Note the large muscovite mica grains exposed on the well defined cleavage surfaces. Middle Mountain, Idaho.

Transverse Cleavage

Transverse cleavage transects known older foliation such as bedding or earlier cleavage; it is generally parallel to the axial planes of folds. In cases where the transverse cleavage is actually observed to be parallel to the axial planes of folds such as at the fold hinge, it can be referred to as **axial-plane cleavage** (Figs. 3-6 and 3-7). Axial-plane cleavage appears to be formed during the folding event and intimately related to the folding. By studying cleavage-bedding relationships, you may be able to determine if a bed is overturned (Fig. 3-8). In overturned limbs, the cleavage dip is less than the bedding dip.

The **axial line** is the intersection between bedding or earlier cleavage and the axial-plane cleavage (Fig. 3-9). These axial lines are parallel to fold axes. This is a very important linear feature to be recorded in the field because in many cases a good example of a fold hinge may not be available. Therefore, the axial line might offer the best value of the fold axis.

Cleavage Refraction and Fanning

Cleavage refraction or fanning occurs where cleavage changes attitude as it crosses a contact between layers of different composition or ductilities (Figs. 3-10 and 3-11).

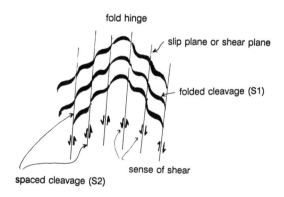

Fig. 3-4. At hinge of folded cleavage, new spaced cleavage begins to form parallel to the axial plane of the folded cleavage. Cleavage formed during the first deformational event (S1) is overprinted by the new spaced cleavage (S2). Mica grains along S1 are deformed into sigmoidal shapes by rotation into new cleavage.

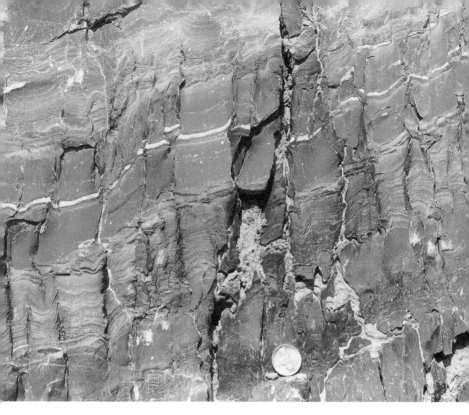

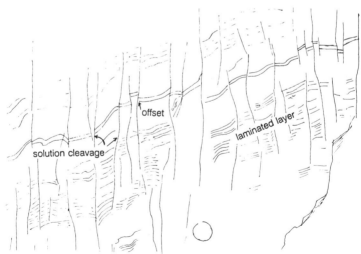

Fig. 3-5. Pressure-solution cleavage in argillite; note the obvious displacement of the bedding planes, particularly the thin white layer; also note the compression folding in the thinly laminated layer. Near Challis, Idaho.

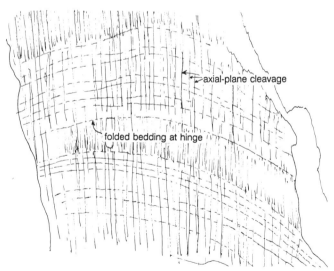

axial-plane cleavage

folded bedding at hinge

Fig. 3-6. Well developed axial-plane cleavage in quartzite at the fold hinge. Near Challis, Idaho.

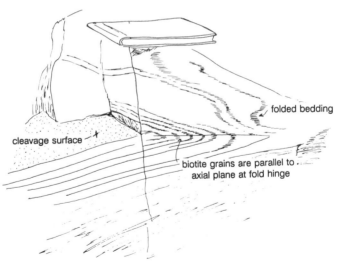

cleavage surface

folded bedding

biotite grains are parallel to
axial plane at fold hinge

Fig. 3-7. Folded quartzite beds defined by thin biotite-rich layers. Mica grains are parallel at the hinge of fold. Note, the flat, smooth cleavage plane. Middle Mountain, Idaho.

Fig. 3-8. Close up of limb of fold in Fig. 3-6 with subvertical axial-plane cleavage intersecting bedding. Near Challis, Idaho.

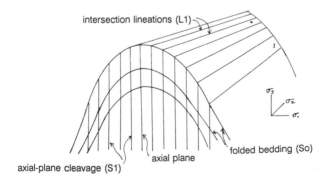

Fig. 3-9. Relationship of folded bedding to axial-plane cleavage and intersection lineations.

Cleavage tends to fan more in competent rocks (sandstones) and less in incompetent rocks (shale). Where strata are folded into alternating beds of sandstone and shale, it is common for the competent sandstone to be cut by fracture cleavage and the incompetent shale to be cut by slaty cleavage. Gradual changes in the lithology of a bed such as with graded bedding may cause curved cleavage planes. Cleavage is generally deflected away from the axial plane in the competent layers and deflected slightly towards the axial plane in the incompetent layers (Fig. 3-10).

Transpositional Layering

Transpositional layering is a common feature of isoclinally folded rocks where the axial-plane foliation is so well developed that it may obscure earlier foliation. Original bedding or earlier cleavage is transformed into parallelism with foliation by folding and ductile shearing during deformation (Fig. 3-12). By this process, spaced cleavage may be converted to transpositional layering by reorientation of platy minerals parallel to the axial plane. When all layering is parallel to the foliation, transposition is complete.

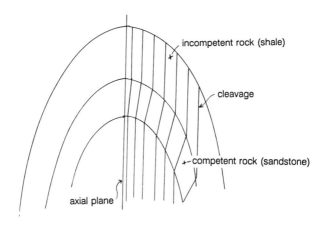

incompetent rock (shale)

cleavage

competent rock (sandstone)

axial plane

Fig. 3-10. Cross section of fold showing refraction of cleavage at the interface or boundary between rocks of contrasting competence. Cleavage tends to parallel the axial plane in the incompetent rock (shale) and fans in the competent rock (sandstone). However, near the axial plane the cleavage in the competent rock tends to parallel the axial plane.

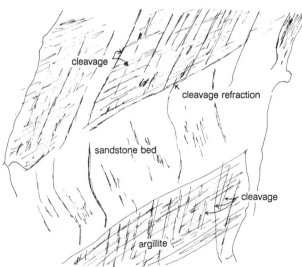

Fig. 3-11. Refraction of cleavage through sandstone bed. Note that the cleavage in sandstone is more poorly developed than it is in the surrounding argillite. Near Challis, Idaho.

65

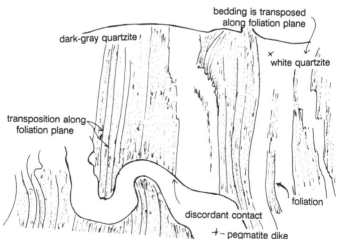

Fig. 3-12. Transposition of quartzite bedding or compositional layering along foliation plane causes lower pegmatite dike to be isoclinally folded. This dike is therefore synkinematic to the deformational event that caused the transposition of the bedding. However, the pegmatite dike at the top of the photo has not been deformed and is therefore postkinematic.

66

Origin of Cleavage

Pressure Solution. Compression perpendicular to cleavage results in shortening through pressure solution of carbonate, silica and feldspar along the cleavage (Fig. 3-5). There is now substantial evidence that a large amount of material may be removed from fine-grained rocks composed of silt and clay during high pressure metamorphism. This evidence includes the density of the metasedimentary rocks, rearrangement of minerals during the compression and compaction and the net loss of soluble minerals such as quartz and calcite. It is very likely that pressure solution, recrystallization and reorientation of minerals all play a role in cleavage formation. Pressure solution is probably dominant in the early stages of cleavage development, and recrystallization may be dominant in the later stages.

The amount of strain varies in different parts of a fold. For example, there is a much greater volume loss from pressure solution in the hinges than the limbs. Evidence for pressure solution includes the following: (1) insoluble minerals such as clay, mica and iron oxides occurring along the cleavage; (2) soluble minerals such as quartz and calcite occurring between the cleavage planes; (3) offset layers that intersect cleavage at an angle; and (4) grains or fossils of soluble minerals truncated at the cleavage.

Shear Foliation without Recrystallization. Cleavage or schistosity may be caused by shearing the rock under high pressure but without recrystallization (Fig. 3-13). Platy mica grains in the influence zone of the shear are dragged out of their original orientation and rotated into parallel alignment

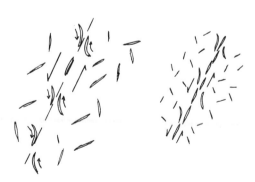

Fig. 3-13. Mica grains are pulled into parallel alignment along shear.

along the shear zone. Another example of shear foliation occurs in crenulation or strain-slip foliation. Cleavage in both cases may be enhanced by crystallization of new micas and recrystallization along the new shear zone.

Ductile Shear Zones. Schistosity may be formed if you shear the rock intensely under high pressure and drag platy minerals into parallelism while they are experiencing recrystallization. This process is important in ductile shear zones.

Growth of Micas Perpendicular to Major Axis of Compression. Cleavage may form by crystallization of mica grains along parallel cleavage planes and perpendicular to the major axis of compression. Axial-plane cleavage is a common example. (Figs. 3-6 and 3-7). The mica plates that existed prior to the deformation are rotated into parallel alignment with the cleavage planes.

Bedding Plane Cleavage or Schistosity. On first observation cleavage or schistosity appears to parallel bedding; however, it only does so if the major compressive force is perpendicular to the bedding. In most cases, cleavage is not related to bedding, but if the two are parallel, the alignment is coincidental. If the bedding is perpendicular or nearly perpendicular to the major compressive force, the mica layers (cleavage) will be very well developed along the beds or layers that were formerly rich in clay minerals; this enhances the appearance of cleavage growing along the bedding plane. With close inspection you may see cleavage transecting bedding at very small angles (Fig. 3-14); however, the best evidence is to find an isoclinal fold hinge (Fig. 3-15). In axial-plane cleavage, mica plates are parallel to the axial plane at the hinge rather than tangential to the folded foliation.

Flexural-Slip Folds. Cleavage could develop along bedding planes from shear due to adjustment between layers (Fig. 3-17). However, hinge areas would not have cleavage because there is very little movement between the beds.

Primary Foliation Distinguished from Secondary Foliation in Granitic Rock

Primary foliation tends to follow contacts parallel to the country rock contact even with abrupt changes in the direction of the contact. Conversely, secondary foliation cuts across contacts even though it commonly parallels contacts of large plutons. If secondary foliation is parallel to a contact, such foliation will continue without deflection even if the contact changes direction (Fig. 3-17).

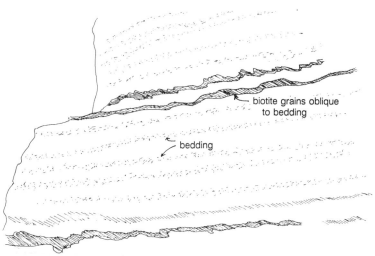

biotite grains oblique
to bedding

bedding

Fig. 3-14. Bedding of quartzite is defined by thin layers of biotite. Note that the grains of biotite are parallel to the axial plane and aligned at an oblique angle to the bedding. Cleavage, which is parallel to the biotite grains, is oblique to the bedding. Middle Mountain, Idaho.

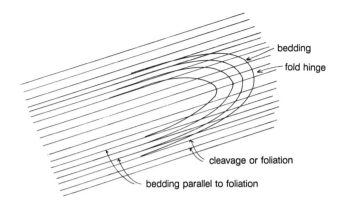

Fig. 3-15. In most high-grade rocks, cleavage or foliation tends to be parallel to bedding except at fold hinges.

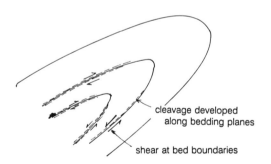

Fig. 3-16. This cross section of a flexural-slip fold shows shearing and relative displacement at the boundaries of beds. Cleavage may develop along the bedding as a consequence of the shearing.

Primary foliation wraps around inclusions; whereas, secondary foliation continues across inclusions. One of the most important criteria for primary foliation is that the strike and dip of the foliation varies greatly over a small area in contrast to the strike and dip of secondary foliation which tends to remain constant over a large area.

Boudinage Structures and Competency Contrasts

Incompetent rocks such as schists will extend more rapidly in the direction of least compressive stress than competent rocks such as quartzites and granite. This length differential causes folding in the incompetent layer and boudinage in the competent layer. Also, frictional drag at the boundaries between layers of different competency contrasts may cause drag folds.

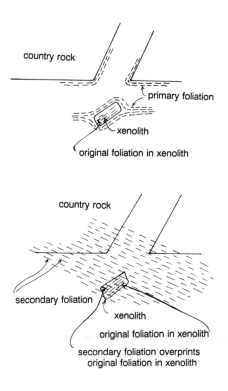

Fig. 3-17. Primary foliation in igneous intrusion wraps around contacts of wall rock and xenolith; whereas, secondary foliation cuts across contacts.

Overprinting of Successive Deformation Events

If there were more than one deformational event, it is important to determine the relative ages of each deformation. Each event may be characterized by its own distinctive S-plane and fold event lineation. The structures of each generation were formed during the same period of time and consequently share the same style. Each successive deformation affects the previous deformation(s). Fold generations are labeled B1, B2, B3 or F1, F2, F3; lineations are labeled L1, L2, L3; and foliations are labeled S1, S2, S3.

Relative Dating of Intrusion to Deformation Event

The determination of whether an intrusive is prekinematic (predates strain), synkinematic (during strain), or postkinematic (postdates strain) can be determined by the field evidence.

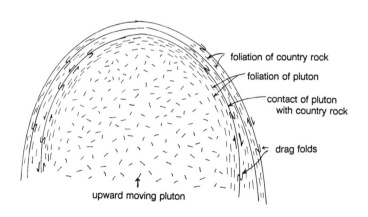

Fig. 3-18. Gneissic border zone caused by the upward movement of a pluton. Small drag folds indicate the relative displacement or direction the pluton moved relative to the country rock. The foliation in the outer margin of the pluton tends to parallel the foliation in the country rock. This foliation, created by the upward-moving pluton, is generally strongest near the contact. In mesozonal plutons, contacts are also characterized by concordant intrusions such as lit-par-lit injection gneiss.

Prekinematic Intrusion. Evidence for prekinematic intrusion includes the following: (1) intrusion predates the effects of the deformation; (2) schistosity cuts across borders of intrusives; and (3) the position or shape of the dikes is not related to structural trends of the deformation.

Synkinematic Intrusion. Evidence for sykinematic intrusion includes the following: (1) cleavage in the country rock parallels that of the intrusion, but cleavage is better developed in the host rock (Fig. 3-18); (2) fold development is weaker in an intrusive than the country rock, but both will have the same axial trends and style; (3) axial-plane dikes and lit-par-lit injection penetrate along planes of cleavage; these dikes or sills are parallel to the cleavage and have the same cleavage as the host rock, but not as strongly developed (Fig. 3-19); and (4) rotated xenoliths of metamorphosed country rock may be found in the intrusion.

Postkinematic Intrusion. Evidence for postkinematic intrusion includes the following: (1) intrusions are free of all deformational features; and (2) intrusions cross cut metamorphic features.

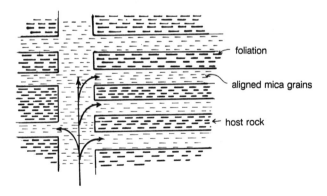

Fig. 3-19. Cross section of lit-par-lit injection gneiss with mica grains in the injected material parallel to the foliation in the host rock. The granitic magma entered as a dike (discordant) and then found the greatest ease of penetration along the foliation planes of the host rock.

4 Lineation

Terminology

Primary lineation is commonly expressed as crystals aligned with the flow direction of the magma; elongate crystals such as hornblende line up to form flow lines.

Secondary lineation (Fig. 4-1) is an alignment of features or mineral grains in response to deformational processes.

Elongation. Conglomerate pebbles, boulders, ooids, orbicules and fossils are flattened or elongated in the plane of cleavage; they are also parallel to the axial plane of folds,

Fig. 4-1. Large-scale, dip-slip lineations on plane of cleavage of micaceous quartzite. At this location, cleavage is parallel to bedding. Middle Mountain, Idaho.

especially in the hinge area. Objects such as ooids, which were originally spheroidal, will become strain ellipsoids after deformation.

Mineral Lineations. Prismatic acicular or elongate crystals such as amphiboles, feldspar, quartz and micas may grow with the long (C) axis in parallel alignment. This preferred orientation is commonly parallel to the axes of related folds.

Fold lineations are generally formed by rotation and the lineations are parallel to the main fold axis. Such folds include flexure folds, passive flow folds, parasitic folds, corrugations, crenulations, fold hinges, drag folds, rootless folds and mullions.

Rods are elongate or cylindrical bodies of quartz, calcite or other mineral, either segregated or introduced in the rock; typically a monominerallic mineral aggregation.

Mullions are long cylindrical or columnar surfaces that may be polished or covered by a film of micas (Fig. 4-2). They may also be defined as the corrugated boundaries between two rock types of contrasting competence. The surface is generally marked by a contact between layers of contrasting ductilities. Mullions are referred to as fold mullions, bedding mullions and cleavage mullions.

Intersection lineations generally occur as one or two types: (1) intersection of cleavage and bedding, particularly axial-plane cleavage at the hinge area of folds (Figs. 4-3, 4-4 and 4-5); and (2) intersection of older cleavage by younger cleavage (Fig. 4-6).

Slippage lineations include slickensides, grooves, striations, fibrons (crystals of calcite, chlorite, quartz and iron oxide), and mineral smears on bedding surfaces or slip cleavages (shears). These lineations are not considered to be penetrative or fabric elements. A common example is slickenside striae on fault surfaces. The long axis is oriented in the movement direction and the surface may be stepped to indicate the relative movement direction along the slip surface.

Strike is the direction or trend of a line formed by the intersection of the planar feature to a horizontal plane.

Dip is the angle between the planar feature and horizontal plane.

Attitude is the three-dimensional orientation of some geologic feature. The attitude of planar features, such as beds or joints, is defined by strike and dip; and the attitude of linear features, such as a fold or boudin axes, is defined by bearing and plunge.

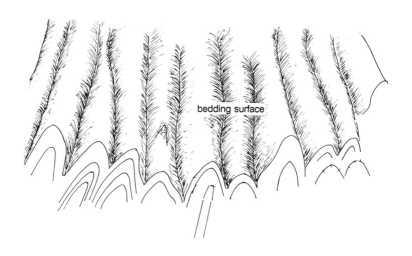

bedding surface

Fig. 4-2. Fold mullions consisting of curved cylindrical surfaces of bedding and cleavage parallel to one another. Middle Mountain, Idaho.

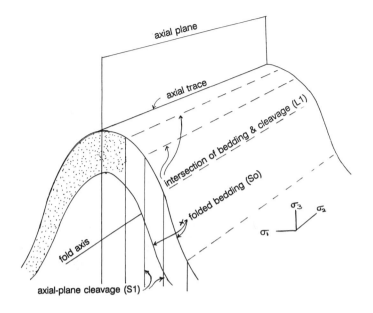

Fig. 4-3. Relationship between bedding and axial-plane cleavage. Lineations are formed by the intersection of cleavage and bedding.

FEATURES OF FOLDED ROCKS

Small folds commonly reflect the style and symmetry of larger folds, and tell us much about the deformation. It is important to observe, sketch and describe the cross section of folds or the view perpendicular to the hinge line.

Fold Terminology

Fold is any bend in a geological surface.

Fold crest is the axial line which is topographically higher than any other on a folded surface.

Fold trough is the axial line which is topographically lower than any other on a folded surface.

Axial plane is the surface connecting the hinges (Fig. 4-7).

Fold wavelength is the distance from anticlinal hinge to anticlinal hinge.

Vergence is the direction an asymmetric fold is overturned.

Hinge is the most strongly curved portion of a fold; there may be an anticlinal and synclinal hinge (Figs. 4-8 and 4-9).

77

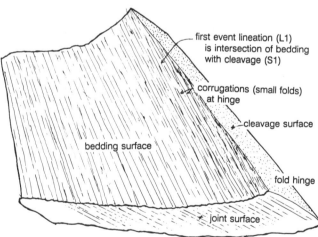

first event lineation (L1)
is intersection of bedding
with cleavage (S1)

corrugations (small folds)
at hinge

cleavage surface

bedding surface

fold hinge

joint surface

Fig. 4-4. This folded bed of quartzite has well developed first event lineations formed by intersection of bedding with cleavage. The bed thickens and lineations are more pronounced towards the fold hinge. Middle Mountain, Idaho.

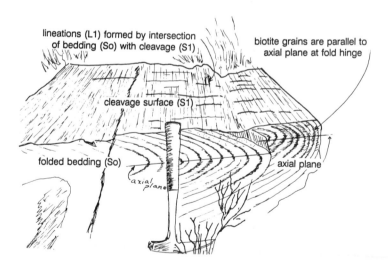

lineations (L1) formed by intersection of bedding (So) with cleavage (S1)

biotite grains are parallel to axial plane at fold hinge

cleavage surface (S1)

folded bedding (So)

axial plane

axial plane

Fig. 4-5. Top surface of fold is axial-plane cleavage in isoclinally folded bedding. Middle Mountain, Idaho.

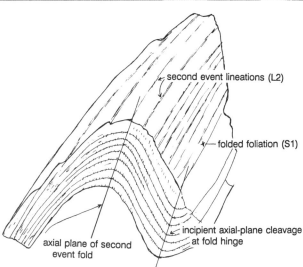

second event lineations (L2)

folded foliation (S1)

incipient axial-plane cleavage
at fold hinge

axial plane of second
event fold

Fig. 4-6. Second generation fold of folded foliation (S1) in schist. The well developed lineations on this folded schistosity (S2) surface are parallel to the axis of this fold and are formed by the intersection of the folded schistosity with a later axial-plane cleavage (S2). Near Shoup, Idaho.

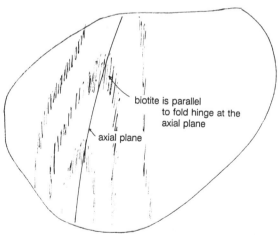

biotite is parallel
to fold hinge at the
axial plane

axial plane

Fig. 4-7. Well rounded pebble from the Big Wood River, near Shoshone, Idaho. Quartzite pebble displays isoclinal fold defined by biotite mica; the mica grains are all aligned parallel with axial-plane cleavage even at the hinge area.

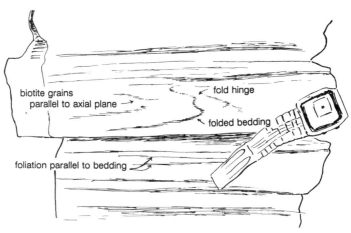

biotite grains
parallel to axial plane →

fold hinge

folded bedding

foliation parallel to bedding

Fig. 4-8. Strongly foliated quartzite by first event. Although generally the foliation parallels bedding, an obscure first event fold hinge would argue against measuring section in such rock because you may be counting the same layer more than once. Note that the alignment of mica grains is parallel to the axial plane at the fold hinge. Middle Mountain, Idaho.

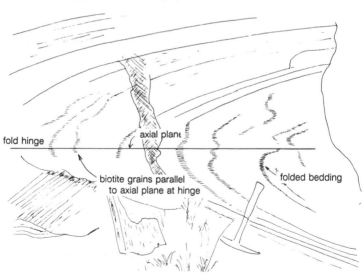

Fig. 4-9. Isoclinally folded bedding in quartzite. Biotite grains are parallel to bedding except at the hinge of the fold. Middle Mountain, Idaho.

Limb is the more weakly curved or uncurved portions of a fold which joins the anticlinal and synclinal hinges.

Reference axes may be used to establish reference directions for the form and symmetry of folds. There are three mutually perpendicular reference axes (a, b and c). The b axis is the fold axis and the a-b plane is the axial plane.

Folding Classified by Type of Deformation (Pressure-Temperature Conditions)

Flexural slip folds occur in the upper levels of the crust where there is little metamorphism. Shear or slip occurs between folded layers with movement parallel to the layers (Fig. 4-10). Evidence for this type of folding includes slickensides on bedding surfaces perpendicular to the fold axis and drag folds parallel to the fold axis. Vergence on drag folds is towards the hinge. There is little or no slip along bedding planes in the hinge area. The layers also tend to be of constant thickness.

Flexural Flow Folds occur at the higher temperature and pressure in the crust than flexural slip folds. There is no discrete surface of slip between layers. Instead, slip is at the granular scale and flow occurs within layers. Flexural flow folds have some layers that deform by ductile flow; whereas, others remain brittle. So high ductility contrasts between layers occur in rocks of low to moderate metamorphic grade. Folds are mostly similar-type, but some may be parallel with thickening in the axial zones and thinning on the limbs.

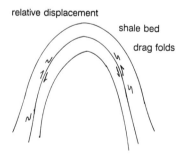

relative displacement

shale bed

drag folds

Fig. 4-10. Flexural-slip fold of shale and sandstone beds with shearing between beds. Relative displacement indicated by axial planes of drag folds developed in the incompetent shale beds.

Passive folds occur in deep and very ductile rocks; cleavage and fabric develop and bedding becomes less important. Folds tend to be of the similar type (Fig. 4-11). There are two types: (1) **passive slip folds** are characterized by slip or shear along cleavage planes, creating a laminar flow where bedding is the marker that records the amount of slip; and (2) **passive flow folds** have no shear or slip along the axial planes. Passive folds reflect uniform ductile flow of the rock unaffected by layering; very little ductility contrast exists between layers and layers are thinned or thickened equally.

Folds Classified According to Axial Plane

Upright folds - the axial plane dips 80 to 90 degrees.
Inclined folds - the axial plane dips 10 to 80 degrees.
Recumbent folds - the axial plane dips less than 10 degrees.

Folds Classified According to Plunge of Fold Axis

Horizontal folds - the axis plunges 0 to 10 degrees.
Plunging folds - the axis plunges 10 to 90 degrees.
Vertical folds - the axis plunges 80 to 90 degrees.

Other Types of Folds

Parallel folds have a constant orthogonal thickness so that successive folded surfaces remain mutually parallel.

Similar folds have successive folded surfaces that are approximately the same shape (Fig. 4-11). Rock flows out of the limbs and towards the hinge area.

Overturned folds are folds with overturned limbs.

Isoclinal folds are folds with parallel limbs.

Reversed folds have the forelimb rotated through the vertical so as to reverse the stratigraphic succession.

Parasitic or second order folds occur on hinge or limbs of larger folds.

Box folds have two hinges with one on either side of a flat fold; the two doubly vergent axial planes create a boxed-shaped fold.

Asymmetric folds have fold vergence in the direction toward which the fold is turned (Figs. 4-12 and 4-13). The fold vergence can be particularly important evidence of the deformation history. The attitude or strike and dip of the axial plane should be recorded.

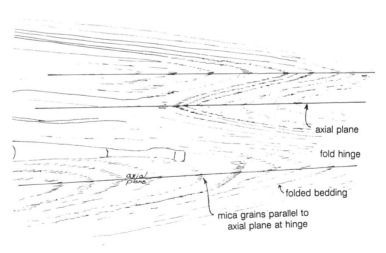

Fig. 4-11. Folded beds in quartzite are defined by layers of biotite mica. Note that the mica grains are parallel throughout the fold hinge. Furthermore, mica is parallel to bedding except at the fold hinge. Middle Mountain, Idaho.

Ptygmatic folds are a common feature of high-grade metamorphic rocks. They typically consist of a single quartzofeldspathic or pegmatitic layer or vein that is convoluted, polyclinal and buckled into tight folds (Fig. 4-13).

Chevron folds have straight or planar limbs and angular hinges.

Kink bands are a type of chevron fold in thinly laminated or foliated rock where the layers are rotated relative to the surrounding layers and the boundaries are marked by angular hinges (Fig 4-14). In **conjugate kink bands** the kinks intersect and occur in related folds with axial surfaces inclined to one another; therefore such folds have conjugate axial planes.

Drag folds are fault-related folds that form where a competent (strong) layer slides past an incompetent (weak) layer. This frictional force causes asymmetric folds to form in the weaker rock in response to the differential movement. To be properly called a drag fold, a fold should have been caused by a fault and have the appropriate vergence. The relative movement direction can be determined by the orientation of the fold axis and the inclination of the axial plane. Drag folds are commonly found in the limbs of flexural slip folds and faults.

Intrafolial folds are characterized by an axial plane that is parallel to the surrounding foliation (Fig. 4-15); the limbs of these typically isoclinal, tight folds are also parallel to the axial plane.

Rootless intrafolial folds are detached from their limbs; this detachment is generally caused by flattening and shearing along the limbs (Figs. 4-16 and 4-17).

Fold mullions are curved cylindrical surfaces of the original bedding or cleavage; mullions are composed of the country rock rather than intruded material such as a dike or sill.

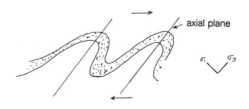

Fig. 4-12. Asymmetric folded layer with clockwise rotation (simple shear) and vergence of the axial plane to the east.

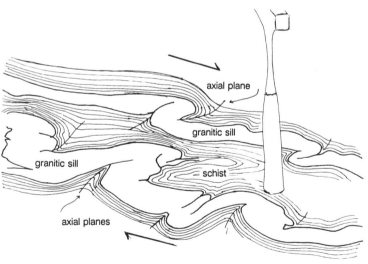

Fig. 4-13. Folded granitic sills concordant to the schistosity of the surrounding schist. The axial plane has vergence to the right. Near Shoup, Idaho.

Anticline is used where rocks get progressively older toward the concave side of the hinge; this type of fold closes upward with older rocks in its core.

Syncline is used where rocks get progressively younger toward the concave side of the hinge; this type of fold closes downward with older rocks in its core.

Where the stratigraphic succession is unknown or not significant, such as in some highly deformed rocks, the terms **antiform** and **synform** are used. An **antiform** is a fold that is convex upward, and a **synform** is a fold that is concave upward.

Evidence of Multiply Deformed Rocks

The following features are evidence for multiply deformed rocks: (1) lineations are curvilinear (Figs. 4-18, 4-19 and 4-20); (2) foliations are folded (Figs. 4-21 and 4-22); (3) multiple foliations intersecting one another (Figs. 4-18, 4-23 and 4-24) (4) multiple fold orientations causes **dome and basin** structure (interference by two or more generations of folds on foliation or bedding surfaces) (Fig. 4-25); (5) tight and isoclinal first generation folds have axial-plane cleavage (Fig. 4-26); (6) later fold generations are more open than earlier generation folds (Figs. 4-27 and 4-28); (7) latest folds have no axial-plane foliation (Figs. 4-27 and 4-28); (8) later folds have more steeply dipping axial surfaces (Fig. 4-28); (9) axial-plane cleavage is folded with a different attitude of fold than folded bedding; (10) new set of folds may overprint old set (4-27 and 4-28); and (11) multiple styles of folds exist in the same area.

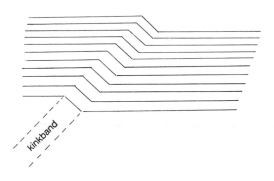

Fig. 4-14. Kink band deflects layers.

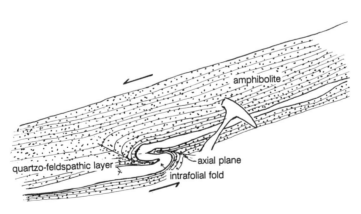

Fig. 4-15. Intrafolial quartzo-feldspathic layer in amphibolite; axial plane of fold is parallel to strong foliation in the amphibolite. An intrafolial fold is an isolated, isoclinal fold lying within the foliation. Near Shoup, Idaho.

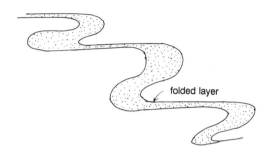

folded layer

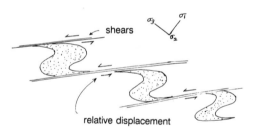

shears

σ_3 σ_1

σ_2

relative displacement

Fig. 4-16. Rootless intrafolial folds caused by shearing along limbs.

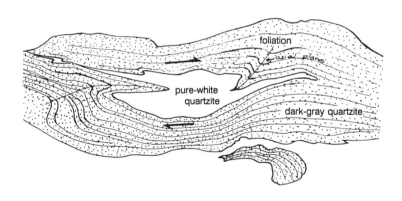

Fig. 4-17. Rootless, intrafolial fold of pure-white quartzite in dark quartzite. This fold is thickened at hinges and detached at limbs by shearing and flattening. Near Shoup, Idaho.

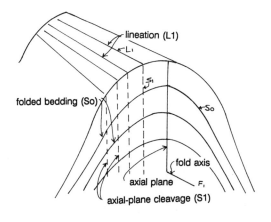

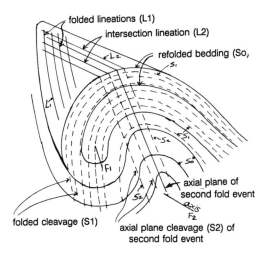

Fig. 4-18. Refolded fold with new axial-plane cleavage (S2) overprinting previous structures. Lineations (L2) are formed by the intersection of the axial-plane cleavage (S2) with the older axial-plane cleavage (S1). The second folding event folded the first event lineations (L1) as well as the first event cleavage (S1).

93

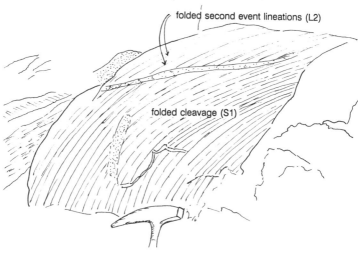

folded second event lineations (L2)

folded cleavage (S1)

Fig. 4-19. Second event lineations on folded schistosity surface formed during the first event. The second event fold axis is not parallel to the first event lineations, but instead, deforms the earlier structures. Near Shoup, Idaho.

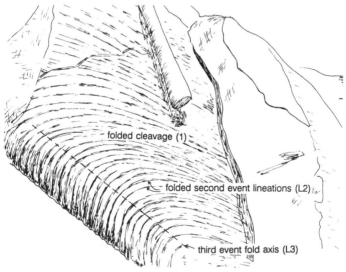

folded cleavage (1)

folded second event lineations (L2)

third event fold axis (L3)

Fig. 4-20. Second event lineations (small folds) on first event foliations are refolded into third event isoclinal fold. Near Shoup, Idaho.

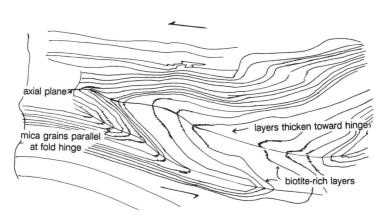

Fig. 4-21. Folded schistosity is defined by thin biotite-rich layers with vergence of axial plane to the left. Mica grains are both tangential to folded schistosity and parallel to the axial plane at the fold hinge. Near Elk City, Idaho.

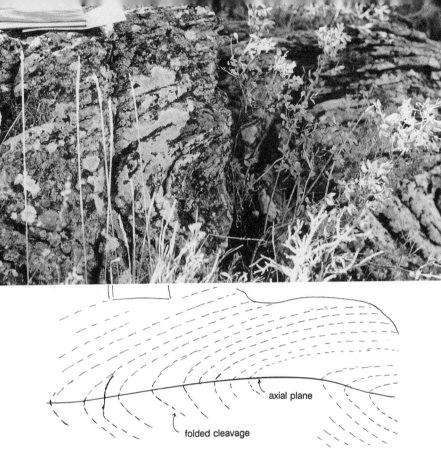

Fig. 4-22. Isoclinally folded schistosity in orthogneiss. Middle Mountain, Idaho.

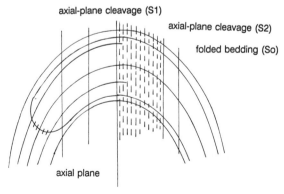

Fig. 4-23. Cross section of folded bedding (So) with axial-plane cleavage (S1) formed during the first folding event. The folded bedding (S0) and axial-plane cleavage (S1) are folded by a second event and a new axial-plane cleavage (S2) is developed.

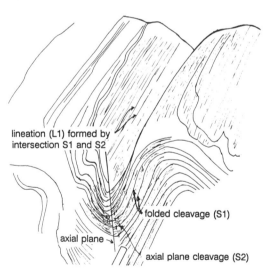

lineation (L1) formed by
intersection S1 and S2

folded cleavage (S1)

axial plane

axial plane cleavage (S2)

Fig. 4-24. Folded foliation (S1) is intersected by axial-plane cleavage which makes lineations (L2) on folded foliation. Mica is both tangential to folded foliation (S1) and parallel to the axial plane (S2) at the fold hinges. Near Shoup, Idaho.

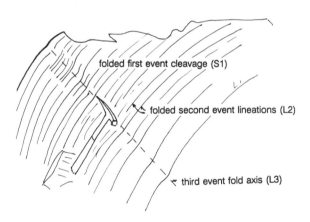

folded first event cleavage (S1)

folded second event lineations (L2)

third event fold axis (L3)

Fig. 4-25. Second event fold axes and lineations are parallel to the hammer handle and are overprinted on the first event foliation surface. The third event folds are perpendicular to the hammer handle. This is an example of dome-and-basin structure. Near Shoup, Idaho.

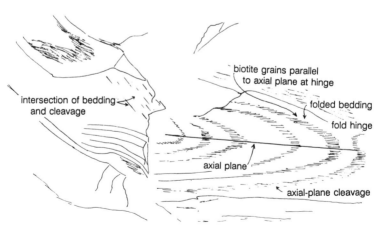

intersection of bedding and cleavage

biotite grains parallel to axial plane at hinge

folded bedding

fold hinge

axial plane

axial-plane cleavage

Fig. 4-26. Hinge of isoclinally folded quartzite is defined by mica-rich layers. Note that the beds are thicker at the hinge than limbs and that the micas are parallel at the hinge. A first event fold near Shoup, Idaho.

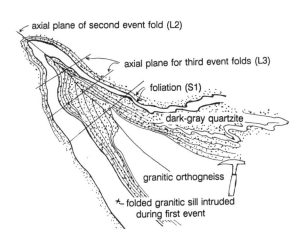

Fig. 4-27. An example of a refolded fold with tightly folded foliation formed during the second event and gentle folds formed during the third event.

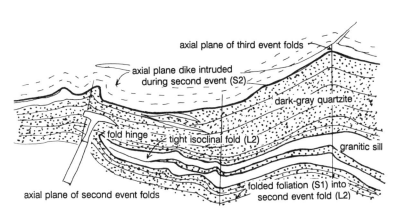

Fig. 4-28. Tightly folded granitic sill, which was intruded during the second event, is concordant to the foliation in the dark-gray quartzite; the structure was folded again into open folds during the third event.

102

BOUDINAGE FEATURES

Many layered rocks are composed of successive layers with contrasting competence; and these rocks of differing competence react differently to deformation. In other words, the competence contrasts give differing rates of strain in two contrasting rock types. Common examples of incompetent rock include shale and schist; whereas, examples of competent rock include granite or quartzite. As a general rule the more silica rich the rock, the more competent it is. At high metamorphic grades, both competent and incompetent rock are ductile; however, there will still be a contrast. At low metamorphic grades, competent layers are brittle and incompetent layers are ductile. Typically, the more competent layer is pulled apart and segmented.

Deformation of successive layers of contrasting competence may cause mullions and boudinage structures, particularly where a competent layer is enclosed within incompetent rock and stretched. Boudins form a series of sausage or cylindrical-shaped forms. They are difficult to identify or determine the orientation because they are commonly seen in cross section or at an oblique angle.

Boudins typically form from beds, dikes or sills which are more competent than the surrounding rocks. The more incompetent layers conform to the shape of the boudin and form folds at the necked or constricted areas. Therefore boudins and their associated folds are developed at the same time and are closely related.

With compression normal to the layering, the less competent layers flow out parallel to the layering, while the more rigid layers fail by tension because of frictional drag on the upper and lower surfaces (Fig. 4-29).

The long axes of the boudins are generally parallel to the fold axes of the deformational event (Fig. 4-30). However, exceptions to this rule are common. Boudins commonly occur along fault zones and are oriented normal to the direction of displacement of the fault.

Rotation of Boudins

Boudins may be rotated on their long axis transforming a basic orthorhombic boudin into a monoclinic one. This rotation may be expressed as sigmoidal (Figs. 4-31 and 4-32) profiles along the cross section of boudins. Several criteria can be used as evidence that rotation had occurred and also to determine

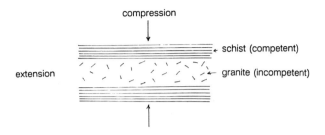

Fig. 4-29. Development of boudins by pure shear with compression perpendicular to the competent layer.

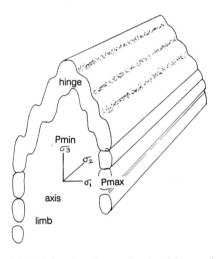

Fig. 4-30. Fold with thickening towards the hinge also displays thinning and boudinage along the limbs.

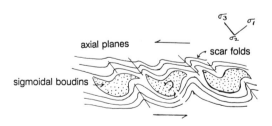

Fig. 4-31. Sigmoidal boudin profiles result from simple shear and low ductility.

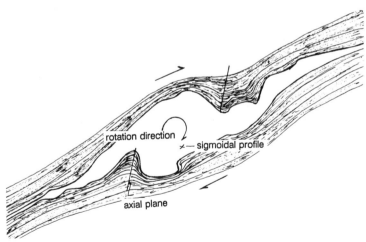

rotation direction

sigmoidal profile

axial plane

Fig. 4-32. Quartzo-feldspathic layer, conformable to surrounding micaceous quartzite, is deformed into sigmoidal profile indicating clockwise rotation. Near Elk City, Idaho.

whether the rotation was clockwise or counter-clockwise: (1) the shape of the boudin profile (Fig. 4-33); (2) the rotation direction of the boudin (Figs. 4-34 and 4-35A & B); and (3) rotated foliation (Fig. 4-36). The vergence direction of folds within the incompetent rock may also indicate the rotation direction.

Chocolate-Block (Tablet) Boudins

Most boudins indicate extension in only one direction; however, if there is extension in two directions, chocolate block or tablet boudins will form. Although uncommon, ellipsoidal boudins do exist and may approximate the strain ellipsoid. Consequently, boudins may represent strain indicators.

Boudin Profiles

Because the competency contrasts that exist between a boudinaged layer and its enclosing matrix is partly a function of temperature, the same two contrasting rock types may produce boudins of entirely different profiles in different pressure-temperature environments. Boudin profiles include sigmoidal (Fig. 4-33), lenticular (Figs. 4-39 and 4-40), rectangular and barrel shapes (Figs. 4-35A and 4-35B). Four factors appear to control the profiles: (1) the competence contrast between the boudinaged layer and its enclosing matrix; (2) the degree of rotation experienced by the boudin; (3) the amount of flattening perpendicular to the layering; and (4) the thickness of the boudinaged layer.

As a general rule you may expect blocky, angular boudin profiles where competency contrasts are large (Figs. 4-35A, 4-35B and 4-41) and lenticular or pinch-and-swell boudins where the contrasts are small (Figs. 4-42, 4-43 and 4-44). Where the competent layer is brittle, extension or tensile joints may form at regular intervals; these joints may be filled by minerals such as quartz and calcite (Fig. 2-6).

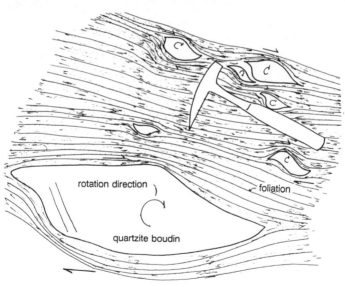

Fig. 4-33. Quartzite boudins in micaceous quartzite have sigmoidal profiles indicating clockwise rotation. Folds in necked areas also indicate clockwise rotation. Near Shoup, Idaho.

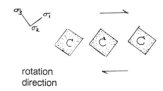

rotation
direction

Fig. 4-34. Rotated blocky boudins occur from simple shear and high ductility contrasts between the competent and incompetent layers of rock.

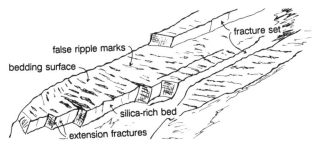

Fig. 4-35A. Fractured blocks with slight rotation in a micaceous quartzite. From a distance the bedding surface appears to have ripple marks. However, these "false" ripple marks are the upper surface of a boudinaged silica-rich bed. The 20-m-thick silica-rich layers are interbedded with more micaceous layers. The extension fractures are formed in the more silica-rich layers. Near Ellis, Idaho.

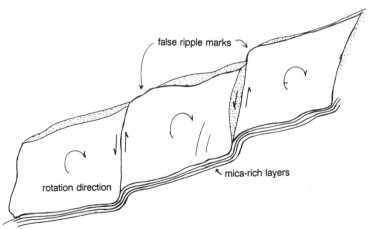

false ripple marks

rotation direction

mica-rich layers

Fig. 4-35B. Close-up of Fig. 4-35A.

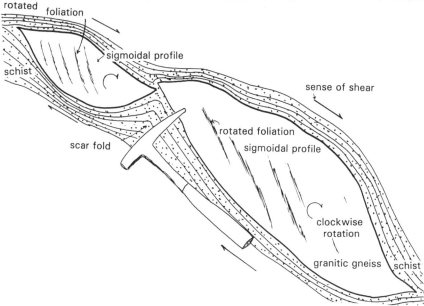

Fig. 4-36. Sill of granitic gneiss is enclosed in less competent micaceous quartzite. The granitic gneiss was deformed into sigmoidal profiles. The sigmoidal profiles and the oblique foliation in the boudins both indicate a clockwise rotation. Near Shoup, Idaho.

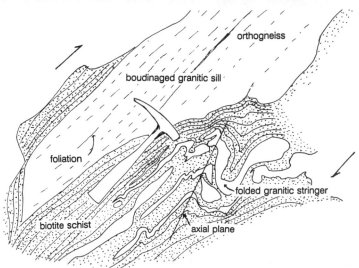

Fig. 4-37. Small folds are ubiquitous in the necked area of boudins. The folds form in the incompetent layer adjacent to the competent boudinaged layer. Vergence direction on the axial plane of this fold indicates clockwise rotation. Near Shoup, Idaho.

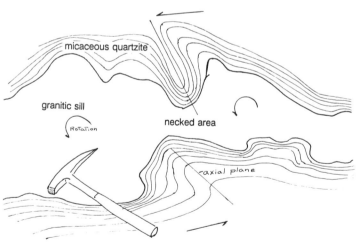

Fig. 4-38. Granitic sill in micaceous quartzite is deformed into boudins. Folds of the less competent micaceous quartzite are formed in the necked areas. Vergence of these folds indicate a counter-clockwise rotation. The axes of these folds parallel the long dimension of the boudins. Near Shoup, Idaho.

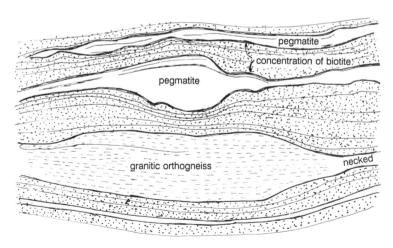

Fig. 4-39. Lenticular boudins of granitic gneiss are conformably enclosed by micaceous quartzite. Strong foliation in the gneiss parallels that in the quartzite. Near Elk City, Idaho.

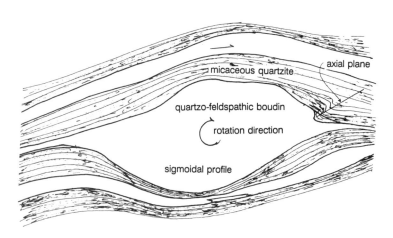

Fig. 4-40. Granitic sill in micaceous quartzite is deformed into lenticular boudins; drag folds in micaceous quartzite give tenuous evidence for clockwise rotation. Near Elk City, Idaho.

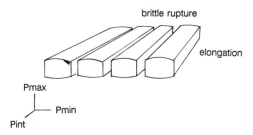

Fig. 4-41. Blocky boudins are developed if there is a high ductility contrast between the competent and incompetent rocks.

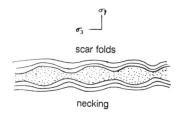

Fig. 4-42. Pinch-and-swell boudins occur if there is a low ductility contrast between the competent and incompetent rocks.

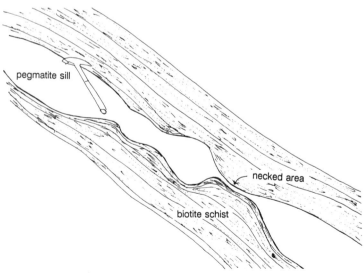

Fig. 4-43. Pegmatite sill in biotite schist is deformed into boudins. Near Shoup, Idaho.

116

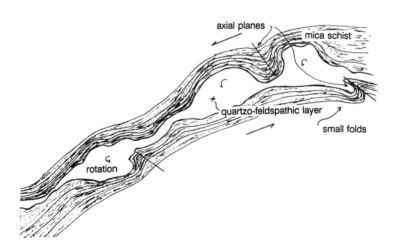

Fig. 4-44. Quartzo-feldspathic layer, conformable to surrounding foliation, is deformed into pinch-and-swell boudins. Near Shoup, Idaho.

117

5 Sedimentary Rocks

Sedimentary rocks are derived from preexisting igneous, sedimentary and metamorphic rocks. These rocks contain many clues of their origin and the conditions that existed while they formed. Sediment, as distinguished from sedimentary rock, is a collective name for loose, solid particles and is generally derived from weathering and erosion of preexisting rock. After formation, sediments are transported by rivers, ocean waves, glaciers, wind or landslides to a basin and deposited. Lithification is the process of converting loose sediment into sedimentary rock and includes the processes of cementation, compaction and crystallization.

Sedimentary rock is formed by lithification of sediments, precipitation from solution, and consolidation of the remains of plants or animals. Coal is an example of sedimentary rock formed from the compression of plant remains.

Rounding of Rock Particles

Rounding occurs during the transportation process by one or more of the erosional agents. Current and wave action in water are particularly effective in causing particles to hit and scrape against one another or a rock surface. The larger the particle, the less distance it needs to travel to become rounded.

Deposition of Sediment

Sorting of sediment by size is also effectively accomplished by moving water. A river sorts sediment by first depositing cobbles, then pebbles, sand, silt and finally clay. The larger the size of sediment, the greater the river's energy necessary to transport it. Deposition is the term used to describe the settling of transported sediment.

Lithification of Clastic Rock

Clastic or detrital sedimentary rock is composed of fragments of preexisting rock. The grains are rounded and sorted during the transportation process. Clastic sediment is generally lithified by cementation. Cementation occurs when material is chemically precipitated in the open spaces of the sediment so as to bind the grains together into a hard rock. Common cements include calcite, silica and iron oxides. A matrix of finer-grained sediments may also partly fill the pore space.

Common Types of Sedimentary Rock

Conglomerate is the coarsest-grained sedimentary rock formed by the cementation of gravel-sized sediments. The gravel is generally rounded; however, it probably did not travel very far. Conglomerates are deposited by running water.

Sandstone is a medium-grained sedimentary rock formed by the cementation of sand-sized sediments, with silt and clay forming the matrix. Sandstones may be deposited by rivers, wind, waves or ocean currents.

Shale is a fine-grained sedimentary rock composed of clay- and silt-sized fragments. Shale is noted for its thin laminations parallel to the bedding. Compaction is very important in the lithification of shales. Before compaction, shale may consist of almost 80 percent water in the pore spaces.

Chemical Sedimentary Rocks are formed by material precipitating from solution. Examples include rock salt, gypsum and limestone.

Organic Sedimentary Rocks consist mostly of the remains of plants and animals. Coal is an organic rock formed from compressed plant remains.

Limestone is a sedimentary rock composed of mostly calcite. Some limestones are chemical precipitates, whereas, others consist mostly of clastic grains of calcite or shells of marine invertebrates. The calcite grains in limestone recrystallize readily so as to form new and larger crystals.

Tillite is a sedimentary rock formed by the lithification of glacial till. **Glacial till** is unsorted and unstratified mixture of clay, silt, sand, gravel, cobbles and boulders deposited by a glacier.

SEDIMENTARY STRUCTURES

The ability to identify and interpret sedimentary structures is crucial for understanding the depositional environment.

Stratification and Bedding

Stratification is the layering in sedimentary rocks defined by changes in mineral composition and texture. In the process of sedimentation, particles of rock or mineral fragments settle out of water according to size, shape, density and velocity of the transporting medium (Fig. 5-1). **Bedding planes** are well-defined surfaces, generally separating the upper and lower surface of two distinctive layers or **beds**. This change may be caused by a period of nondeposition. After a period of non-

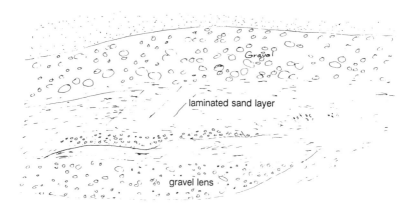

Fig. 5-1. Unconsolidated fluvial deposit with lens of gravel interbedded with layers of sand. The upper sand layer is massive but the middle layer is laminated with cross bedding exposed in the fine pebble layers indicating current from left to right. Near Challis, Idaho.

deposition, it is very likely that the new deposit will differ even slightly from the bed below. A **bed** is a layer of strata thicker than 1 cm which is characterized by uniform sedimentation or a layer of sediment deposited during nearly constant physical conditions. **Laminae** are strata thinner than 1 cm. **Laminations** are caused by relatively brief and minor changes in physical conditions during sedimentation.

Massive beds appear to have no internal structure.

Sorting

Sorting refers to similarity of particle size in a sedimentary rock. Poorly sorted rocks have a great range in particle size; whereas, well sorted rocks have a similar grain size. Wind-blown deposits are generally better sorted than stream deposits. Deposits with poor sorting such as conglomerates are generally transported only a short distance under fairly violent conditions and then deposited rapidly (Fig. 5-2).

Shape of Grains

The more rounded the grain, the greater the amount of transportation or time it has spent in an abrading environment. However, the larger the grain size the less time or activity it takes to become rounded. For example, large boulders may be rounded by only 10 km of downstream movement under flood conditions. Minerals also vary in their hardness and stability to weathering action. Hard and stable minerals such as quartz tend to be enriched in sedimentary deposits that have been transported or reworked over long periods.

Cross Bedding

Cross bedding is characteristic of sand dunes, river deltas and stream channel deposits. It is most typically observed where sedimentary layers or laminations accumulate at a steep angle to the horizontal. Cross bedding forms from the downstream migration of ripples, bedforms and sand waves when particles move over the sloss side and deposit on the lee side. They may form along the front of a growing delta or migrating point bar. Cross bedding is a good indication or wind or current direction. Erosion of the topset beds and truncation of the foreset beds is a good top and bottom indicator (Fig. 5-3). Foresets are either planar (angular) (Fig. 5-4) or tangential (Figs. 5-5, 5-6 and 5-7), depending on how they intersect the lower set boundary. Cross bedding is particularly common in

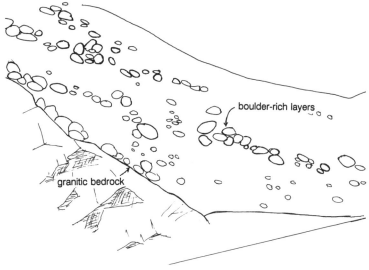

boulder-rich layers

granitic bedrock

Fig. 5-2. Poorly sorted, crudely stratified fluvial deposit of cobbles and boulders overlying granitic bedrock some 12 m above the current level of the Salmon River, Idaho. Note the boulder-rich layers within the deposit.

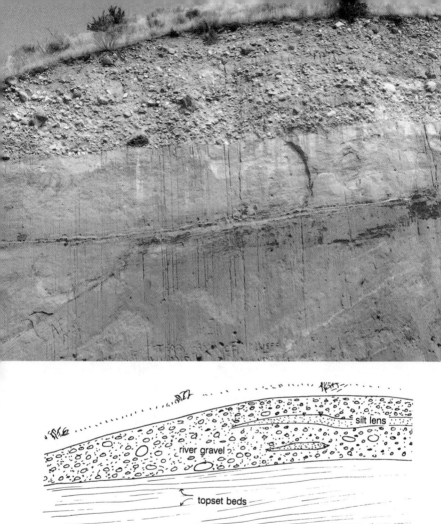

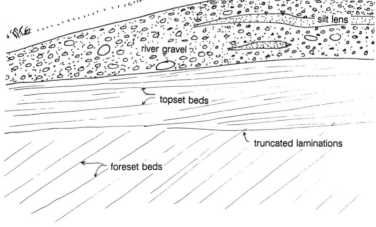

Fig. 5-3. Sandstone with steeply inclined foreset beds and subhorizontal topset beds. Bedding in the pebble conglomerate is parallel to the underlying topset beds of the sandstone. Note the thin layers of sandstone in the conglomerate that help define the bedding. Near Boise, Idaho.

123

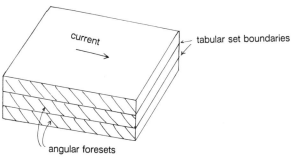

Fig. 5-4. Zion National Park is known for beautifully defined cross beds in the Jurassic Navajo Sandstone. The Navajo desert covered an area of about 390,000 square kilometers reaching from Wyoming to Southern California. Tabular sets of cross bedding with angular foresets are well exposed in the photograph. The diagonal lines in the photograph represent an of advance of dunes across a lowland desert by wind blowing from right to left.

124

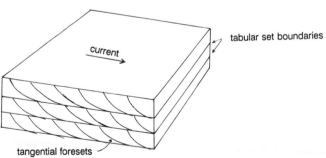

Fig. 5-5. Tabular sets of cross bedding with tangential foresets in the Navajo Sandstone, Zion National Park. The diagonal lines in the photograph represent the advance of dunes by wind blowing from right to left.

Fig. 5-6. Well developed and preserved tangential cross beds in high-grade Precambrian quartzite. Note that the laminations are truncated on the upper surface of the cross beds and tangential on the lower surface. Middle Mountain, Idaho.

beds with abundant sand-sized materials (Figs. 5-8 and 5-9). Cross bedding produced by wind typically forms much thicker sets and the foresets dip at greater angles than cross bedding formed by water.

Ripple Marks

Ripple marks are undulations on a sand or silt surface caused by the movement of wind or water across it (Fig. 5-10). The ridges form at right angles to the direction of the wind or water current. Those formed by stream currents tend to be asymmetrical in profile (Fig. 5-11); and those caused by the oscillation of waves are symmetrical (Fig. 5-12). Current ripple marks have steeper sides in the downcurrent direction and more gradual slopes on the upcurrent side. Where asymmetrical ripple marks are preserved in rocks, it is possible to determine the direction of movement of the wind or water (Fig. 5-13).

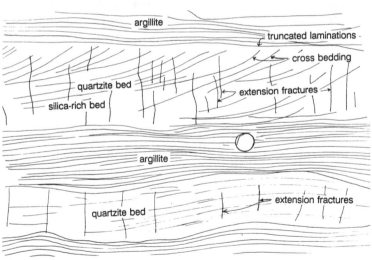

Fig. 5-7. Light, quartz-rich beds interbedded with darker clay-rich beds. The light beds have distinct cross bedding laminations with truncated upper surface and tangential lower surface. The current direction is right to left. The light silica-rich layer has extension fractures perpendicular to the layering because it is more competent than the clay-rich layer. Glacier National Park.

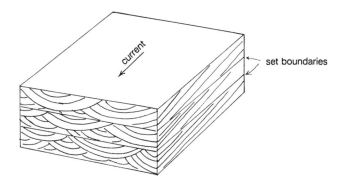

Fig. 5-8. Three tabular sets of cross bedding with trough sets.

Ripple marks are deposited from the traction load or the material carried along the bottom by sliding, rolling and skipping. The size and shape of ripple marks are a function of the current and grain size. **Climbing ripples** display sediment deposition over the entire bed. **Oscillation ripples** are generally symmetrical ripples caused by waves rather than current. Because they have sharp peaks and rounded troughs, they are a good top and bottom indicator.

Sole Marks

Sole marks are structures preserved on the lower side of beds that stand out in positive relief (Fig. 5-14). The depressions are formed by (1) the action of current on the surface of the mud, (2) unequal loading over soft mud, and (3) organisms. They generally occur where siltstone or sandstone overlie ˉmudstone. These structures are not particularly common because they require first a mud or clay stone to be deposited; then the surface is eroded or marked and covered by silt or sandstone. Finally the clay is removed leaving only the sandstone with the sole marks. Sole marks are commonly produced by turbidity currents.

Scour casts are best made in clay-rich mud which later becomes filled by sand or gravel (Fig. 5-15). These current marks are either formed by a tool driven by the current or the current itself. Casts are useful to determine paleocurrent direction. **Flute casts** are the narrow ends of scour casts. **Tool marks** occur where objects carried by the current strike the bottom and make marks parallel to the current direction. Tools

Fig. 5-9. Fluvial deposits exhibiting cut-and-fill structure with well developed trough cross bedding. This coarse sand- and granule-sized material shows well defined laminations. Note that the laminations are tangential to the lower surface, but truncated by the stream scouring on the upper surface. Southwest Idaho.

Fig. 5-10. Well preserved ripple marks on subvertical bedding surface of Precambrian argillite of the Belt supergroup. Northern Idaho.

include sticks, shells, pebbles or rock fragments that are moved along the bottom by current action and makes **groove casts** and **prod casts**. **Groove casts** fill linear grooves formed by current dragging sharp objects along the bottom. Groove casts tend to be straight, parallel and continuous; whereas, **prod casts** are relatively shorter and asymmetrical with the downstream end wider and deeper than the upstream end.

Graded Bedding

Graded bedding occurs in beds where there is a gradation in size of particles, from coarse at the bottom to fine at the top of a layer. Graded beds are deposited where the current slows. In some beds, grain size grades from relatively coarse at the bottom to a finer size at the top. For example, a lithified sequence may grade from a conglomerate at the base to a shale at the top. This is a standard top and bottom indicator, although some exceptions exist. These deposits are called

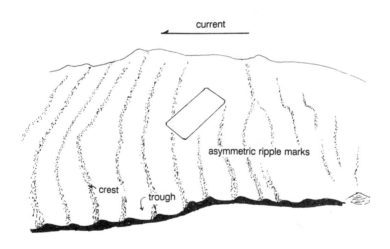

current

asymmetric ripple marks

crest

trough

Fig. 5-11. Asymmetric ripple marks on bedding surface of Precambrian quartzite north of Salmon, Idaho. The asymmetry of the ripples indicates current from right to left.

Fig. 5-12. Cross section of asymmetrical and symmetrical ripple marks. The asymmetrical ripple marks indicate the current direction by their profile.

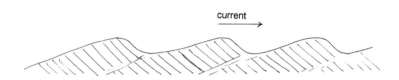

Fig. 5-13. Cross section of asymmetrical ripple marks showing internal structure and how ripple marks migrate by growth on their lee side.

Fig. 5-14. Sole marks of current structures with plant debris on Precambrian bedding surface.

132

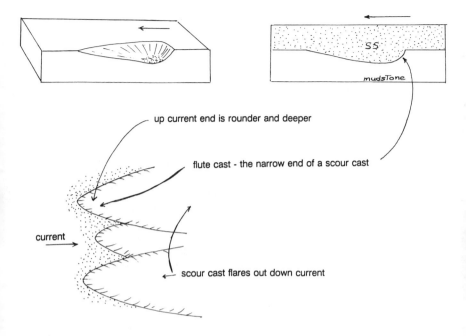

Fig. 5-15. Scour casts form from current eroding a mud surface with subsequent deposition of sand to preserve the feature.

turbidites. They commonly settle in basins that have long and relatively steep slopes. They may settle from dense suspension of sediment in water caused by mud slumps, submarine slides and flood streams. Many air-fall tuffs are graded and can be developed in air or water. From a distance, grading may be apparent from the well defined sharpness of the bottom side of the sandstone bed. The upper surface is poorly defined because sand-sized material grades into the silt- and clay-sized material.

Imbricate Structures

Imbricate structures may develop in streams with strong current action. In such cases, elongate pebbles and cobbles are oriented with the current and dip in the upstream direction (Figs. 5-16 and 5-17).

Stromatolitic Bedding

Stromatolitic bedding in limestone is unlike current bedding. This growth bedding, generally in the form of laminations (Fig. 5-18), is caused by the formation of mats of

Fig. 5-16. Well developed imbricate structure in recently deposited gravel bar in the Salmon River north of Riggins, Idaho. The current flow is from left to right. The long dimension of the pebbles dips steeply in the upstream direction.

bluegreen algae; the laminations may be difficult to identify as stromatolitic in origin if they are not arched in columns. Of course all banded limestone structures are not stromatolites; they can be precipitates such as travertine.

Mud Cracks

Mud cracks are shrinkage or desiccation cracks that form in silt and clay (Figs. 5-19, 5-20 and 5-21). The cracks intersect to form polygons of dried mud layers. When dried each polygon tends to curl concave upwards so that cracks are wider at the surface. They are a good top and bottom indicator. Mud cracks are commonly formed in the intertidal zone where the mud is exposed to the sun during low tide and desiccates. When the tide comes in, these mud cracks may be buried by wet sand and preserved.

Rain Drop Imprints

Rain drop imprints appear as small, circular, shallow pits caused by hail or rain falling on soft mud (Fig. 5-22). They typically have a random distribution and can be used as top and bottom indicators.

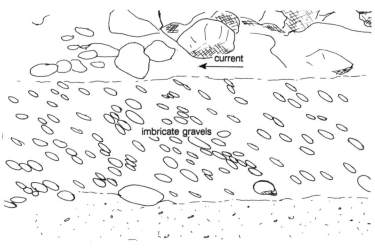

Fig. 5-17. Fluvial deposit with 1-m-thick layer of gravels showing good imbricate structure. The long dimension of the pebbles dips to the right indicating they were deposited in a current flowing from right to left.

Fig. 5-18. Precambrian stromatolitic limestone showing several columns. The laminations are formed organically by deposition of mats of blue-green algae. Belt Supergroup in Glacier National Park.

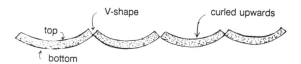

Fig. 5-19. Mud cracks are formed by desiccation and shrinkage of the upper layer of mud during high tide or recent emergence from water.

Fig. 5-20. Precambrian quartzite north of Challis, Idaho showing well-preserved mud cracks on a steeply dipping bedding surface. The mud cracks appear to be significantly elongated in a direction parallel to the dip of the bedding.

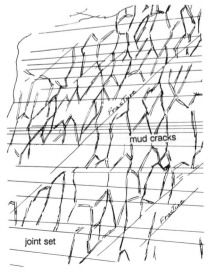

137

Fig. 5-21. Mud crack casts in Belt Supergroup rocks in Glacier National Park.

Fig. 5-22. Rain drop imprint forms a small crater.

Fossil Evidence

Fossils tend to occur along bedding surfaces with the convex side up and can be used as a top and bottom indicator (Fig. 5-23). It is important to determine if fossils are in growth position or have been moved by current. Fossils in reefs tend to be in growth position. Where fossils are in growth position, you should determine the relative abundance of each type. the following evidence indicates that fossils accumulated by current: (1) sharp scour bases; (2) not in growth position; (3) sorting by size or shape; (4) broken or abraded forms; and (5) long axes oriented parallel to the current.

Trace fossils give important clues to depositional environments. These traces are made by organisms touching bottom, moving along bottom, feeding in the sediments and excavating a living space. **Bioturbation** is caused by animal movements that destroy the primary sedimentary fabrics and structures. The amount of disturbance depends on the number of animals, the types of animals, the rate of sediment accumulation and the food content of the sediment. A lack of bioturbation may indicate a lack of oxygen.

Soft Sediment Deformation (Penecontemporaneous Deformation)

Soft sedimentary deformation or **penecontemporaneous deformation** is the deformation that occurs before lithification of the rock. This deformation is a nontectonic type of deformation. Soft sedimentary deformation structures include (1) **slump structures**, (2) **deformed or folded layers**, and (3) **load structures**.

Slump structures occur where sediment deposited on the slope tends to slide downslope (Fig. 5-24).

Deformed or folded layers are sandwiched between undeformed layers (Fig. 5-25).

Load structures are generally caused where sand is deposited on water saturated mud and the weight of new overlying sediment forces interstitial water out of the underlying sediments. Compaction of underlying sediment results in depressions on the upper surface (Fig. 5-26). Load structures are common in deep water turbidites or where there is contrast in grain size.

Diagenetic Structures

Diagenetic structures are secondary structural features in sedimentary rock in which chemical processes redistribute

Fig. 5-23. Subvertical bedding in the Jurassic Nugget Sandstone. Tracks of Jurassic-age creature on bedding surface of sandstone. East of Bear Lake, Idaho.

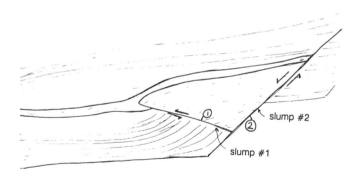

Fig. 5-24. Slump structures with faulting on dark laminated layer in Belt Supergroup. This slump occurred before consolidation of the rock. Glacier National Park.

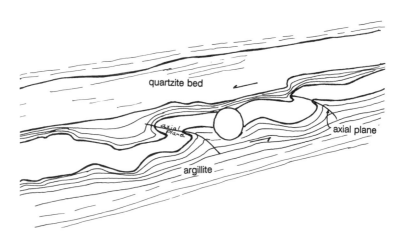

Fig. 5-25. Slump structures with folding and vergence of axial planes to left indicating that upper layers slid right to left over lower layers. The laminations above and below the light folded layers are not disrupted by folding which is typical of gravity slump or downslope sliding of unconsolidated sediments. Glacier National Park.

142

Fig. 5-26. Pillow structures representing possible load casts or algal structures where the laminae are convex upward in Precambrian Belt Supergroup rocks.

soluble material within the sedimentary rock. These structures include (1) **styolites**, (2) **Liesegang banding**, and (3) **compaction folds**.

Styolites are irregular seams caused by solution along surfaces that are approximately perpendicular to the compression. Pressure solution compaction of limestone produces sutured or irregular seams consisting of clay or other insoluble material. If a styolite crosses a fossil, a portion of the fossil will be dissolved. Up to 40 percent of the rock may be dissolved along a seam that contains the insoluble residue.

Liesegang banding occurs as color banding that transects bedding in sedimentary rocks. It is caused by diffusion of iron oxide in ground water.

Compaction folds are formed after deposition. Compaction leads to expulsion of interstitial water and rearrangement of grains of overlying sediments. Compaction orients platy particles such as clay.

Nodules or Concretions

A **concretion** is a hard aggregate of mineral materal formed by precipitation about a nucleus such as a fossil or mineral grain in the pore space of a sedimentary rock. Concretions exist in a great variety of shapes and sizes; however, they are typically spheroidal in form and less than 1 m in diameter. Minerals such as calcite, dolomite, siderate, chert and pyrite are common cementing materials. Most concretions are formed during diagenesis in limestone and shale. Chert nodules are common in limestone, whereas, calcite and dolomite concretions tend to occur in sandstone.

Unconformities

An **unconformity** is a buried surface of erosion separating two rock masses. The term implies a long period of erosion before deposition of the younger rock. Types of unconformities include (1) **disconformity** or **parallel unconformity**, (2) **angular unconformity**, (3) **nonconformity**, and (4) **onlap unconformity**.

Disconformity or parallel unconformity. Strata on both sides of the unconformity are parallel. This type of unconformity is difficult to identify because the contact may look like a bedding plane. Evidence for a disconformity includes truncated fractures, root tubes and dikes as well as soil profiles.

Angular Unconformity. Tilted sedimentary rocks are overlain by younger and more horizontal strata (Fig. 5-27).

Nonconformity. An erosional surface separating older metamorphic or igneous rock from younger sedimentary strata.

Onlap Unconformity. Younger strata appear truncated by unconformity.

Chemical Weathering

Coarse grained sedimentary rocks with sufficient clay and feldspar will develop spheroidal weathering (Figs. 5-28, 5-29 and 5-30). Spheroidal weathering is less common in sedimentary rocks than in igneous rocks.

Rock Colors

Perhaps the most apparent feature of rocks to the observer is the coloration. Although most rocks have a rather drab appearance, some have very distinctive and, in some cases, beautiful colors. Shades of red, green, gray and brown may be caused by iron-bearing minerals. Very light-colored rocks are generally lacking in iron-bearing minerals. The coloration of

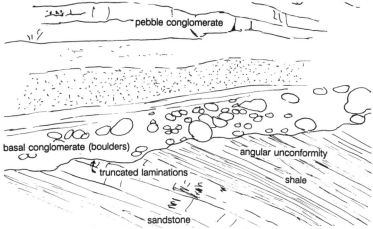

Fig. 5-27. Angular unconformity in Cretaceous sedimentary rocks in central Oregon west of Bend. The old erosional surface is formed on tilted interbedded sandstone and shale layers. Overlying the erosional surface is a basal conglomerate with material up to boulder size.

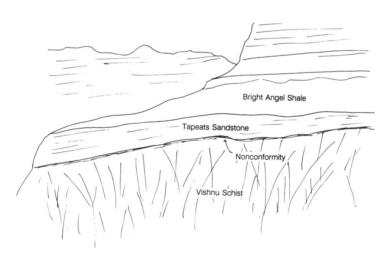

Bright Angel Shale

Tapeats Sandstone

Nonconformity

Vishnu Schist

Photograph of the north wall of the Grand Canyon with excellent example of a nonconformity. The erosional surface is developed on the Precambrian Vishnu Schist. The Cambrian Tapeats Sandstone unconformably overlies the Vishnu Schist.

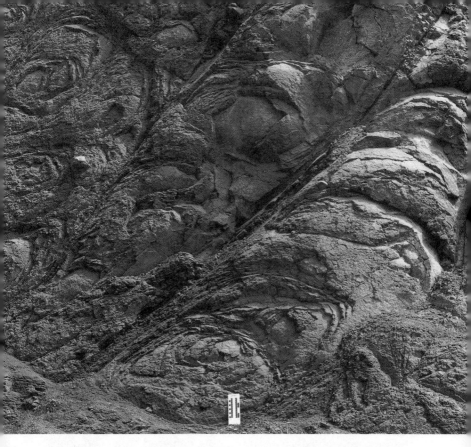

Fig. 5-28. Spheroidal weathering in Cretaceous conglomerate. A set of fractures with spacing of about 1 m appear to control and limit the maximum size of the largest shells. Each successive layer is parallel and about 1 cm thick. East of Bend, Oregon.

sedimentary rocks reflects the environmental conditions that existed during deposition.

Purple and Red Rocks. Purple, maroon and red rocks are stained by the mineral hematite (iron oxide). Hematite results from the decomposition and oxidation of iron-rich minerals such as magnetite, ilmenite, biotite, hornblende and augite. A rock composed of only several percent hematite may be stained a deep red.

Green Rocks. Green sedimentary rocks are typically formed in a reducing environment where oxygen is not available. For sedimentary rocks, this would normally mean deposition in deeper water than red rocks. In a reducing environment, iron combines with silica compounds to form iron silicate minerals.

Fig. 5-29. Exceptionally well developed layers or shells are arranged concentrically about a less weathered core. The shells are successively more weathered towards the fracture where the weathering process began. East of Bend, Oregon.

Then low-grade metamorphism converts the iron silicates to the green mineral chlorite. Chlorite in sedimentary rocks indicates a deep-water depositional environment. Where chlorite-rich strata alternate with hematite-rich strata, a change in sea level probably occurred.

Black Rocks. Higher-grade metamorphism (high heat and pressure) will convert the hematite in red rocks and the chlorite in green rocks to the black minerals magnetite and biotite. An abundance of these minerals will yield a gray to dark gray mineral. Traces of black organic matter will also darken a rock to gray or dark gray.

Weathered Surfaces. Many rocks have a different color on the weathered surface than on a fresh break. Weathering of disseminated pyrite (iron sulfide) in rocks will convert the pyrite to brown or yellow iron hydroxide and iron sulfate.

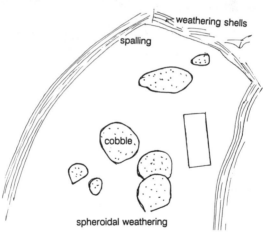

Fig. 5-30. Spheroidal weathering generally leaves the weathered surface smooth and rounded even where the rock has components that have different texture and hardness. The cobbles of the conglomerate are much harder than the surrounding matrix. Where the harder cobble projects out from the matrix, the exposed or projecting portion of the cobble expands more readily than the embedded portion that is constrained by the confining matrix. Consequently, a crack may form along the surface leaving a smooth outer surface despite the variable hardness or durability of the exposed rock. East of Bend, Oregon.

Alluvial fans form a **bajada** in Death Valley California. A bajada is a broad, gently sloping depositional surface formed by coalescing alluvial fans.

Alluvial Fans

Alluvial fans may form in a variety of geologic environments; however, the most commonly described deposits are those that occur in the semiarid Basin and Range Province in the western United States. These fans are primarily formed by infrequent rainstorms in the ranges. Because the slopes of the mountains have very little vegetation or soil amenable to collecting or absorbing the sudden precipitation, water runs off the slopes carrying loose clasts down to the canyon bottom. This loose material is transported through the canyon by a flash flood. When the sediment-charged water emerges from the canyon as a mud flow or debris flow, it spreads out and runs in broad

Longitudinal view (parallel to the channels) showing a sequence of debris-flow deposits with an average thickness of 0.5 m. Note the angular clasts and the poor sorting in each debris-flow deposit. Death Valley, California.

channels following the lowest topography on the alluvial fan.

In plan view the fans are built up by many radiating channel deposits. A large fan is composed of many layers of debris-flow deposits ranging from several centimeters to 3 m thick. Sediment is transported in streams, mud flows and debris flows depending on the ratio of available water to sedimentary material. As a rule, each deposit decreases in thickness from the mouth of the canyon to the outer margin of the fan. Clasts in the debris flows tend to subangular and progressively decrease in size from the mouth of the canyon to the margin of the fan.

Belt Supergroup Sedimentary Rocks

Belt Basin. Sedimentary rocks of the Belt Supergroup formations range from about 850 to 1,450 million years old. In the northwestern United States and in adjacent parts of Canada, these Precambrian rocks occupy the so-called "Belt basin," an area that covers most of the Idaho panhandle as well as adjacent parts of eastern Washington, western Montana and southern Canada. The Belt basin is not strictly a basin but is rather the only re-entrant of a sea that existed off the west coast of North America. Because much of the Belt basin is now covered by younger rocks, the outer limits of the basin cannot be accurately established.

Rock Types. The sedimentary Belt rocks are fine-grained, predominately of clay and silt size, with only a very small percent of the sand-sized component. The rock types include quartzite, dolomite and limestone. Fossil algal forms are found in some of the carbonate rocks. These forms, called stromatolites, represent the only evidence of life in the Belt rocks.

Age of the Belt Rocks. The age of the Belt Supergroup ranges from 1,450 to 850 million years. Belt rocks accumulated in shallow continental basin and adjoining continental margin. They were deposited on crystalline rocks of a continental crust ranging in age from 2,760 to 1,440 million years. During a 600 million year period, a maximum of 20 km of sediments were deposited.

Shallow Water Deposition Environments. Sediments collected in the Belt basin exhibit substantial evidence supporting a shallow water environment for the Belt Supergroup. This evidence includes sedimentary structures such as fossil stromatolitic algae (Fig. 5-18), mud cracks (Fig. 5-21), small-scale cross bedding (Fig. 5-7), ripple marks (Fig. 5-10), salt crystal casts, flute casts, groove casts, load casts (Fig. 5-26) and mud-chip breccia (Fig. 31). Mud-chip breccia consisting of thin layers of dried mud ripped up by moving currents, then redeposited and buried in new sand and silt. All of these features are indicative of a shallow water environment of deposition in or near the tidal zone.

There was a prolonged gradual subsidence which kept pace with sedimentation. Most of the sedimentation took place on flood plains, tidal flats and as bank deposits in shallow marine water. Before metamorphism, the Belt rock types were clays, silts, sands and limes. These sediments were lithified by compaction and natural cementation into shales, siltstones,

Fig. 5-31. Mud chip breccia in Precambrian Belt Supergroup rocks. After desiccation of the mud surface in the intertidal zone, the upper thin layer of mud dries, cracks and forms mud curls. The incoming tide brings sand which buries and preserves the mud chips. This process may occur time after time. Also note the clay balls buried in the sand. Glacier National Park.

sandstones and limestones. Metamorphism converted the rocks into argillites, siltites, quartzites and dolomites.

Stromatolites. Fossil forms of ancient algae or former marine plant life are called stromatolites. These stromatolites are similar to the modern blue-green algae; and in the outcrop they consist of alternating bands of light and dark-colored mineral matter arranged in swirling patterns. Fossilized algae have nearly spherical or ellipsoidal structures ranging from the size of a football to a large reef.

All of the algal forms are collectively called stromatolites or, when forming massive rock, they are called stromatoliths. The genus name for many of the forms is **Collenia**. **Collenia** lived in protected intertidal flats of the ancient Precambrian Beltian sea.

Sand Dunes

Wind-blown sand is typically found in deserts and deposited along shorelines. These deposits are composed of well sorted and well rounded sand grains that may be frosted by impact against one another. The most notable sedimentary structures are large-scale cross bedding with steeply dipping cross beds. When wind looses its velocity and its ability to transport the sand, it is deposited on the ground. Sand deposits tend to assume recognizable shapes (Figs. 5-32 and 5-33). Wind forms sand grains into mounds and ridges called dunes, ranging from less than a meter to more than a hundred meters in height. Some dunes migrate slowly in the direction of the wind. A sand dune acts as a barrier to the wind by creating a wind shadow. This disruption of the flow of air may cause the continued deposition of sand. A cross section or profile of a dune in the direction of blowing wind shows a gentle slope facing the wind and a steep slope to the leeward. A wind shadow exists in front of the leeward slope which causes the wind velocity to decrease. The wind blows the sand grains up the gentle slope and deposits them on the steep leeward slope.

Bruneau Sand Dunes. The Bruneau Sand Dunes State Park, established in 1970, is located 29 km south of Mountain Home, Idaho. Although there are many small dunes in the area, two large, light-gray overlapping sand dunes cover approximately 600 acres. These two imposing dunes are striking, particularly because they dwarf most of the nearby land features. The westernmost dune is reported to be the largest single sand dune in North America (Fig. 5-34), standing about 160 m above the level of a nearby lake.

The existence of these dunes is attributed to the constant wind blowing sand from the southwest. As the wind loses velocity over the basin, sand is deposited on the dunes. Sand has been collecting in the area for more than 30,000 years.

Bedrock Erosional Features of the Big Wood River

The Big Wood River has one of the best examples in the nation of bedrock erosional features (Maley and Oberlindacher, 1993; Maley and Oberlindacher, 1994). The Big Wood River, which is fed from the mountains north of the Snake River Plain, cuts through 0.8 m.y. old basalt in an area north of Shoshone, Idaho. Approximately 10,000 years ago, nearby Black Butte shield volcano erupted basaltic lava which rerouted the Big Wood River. High flows of water from the

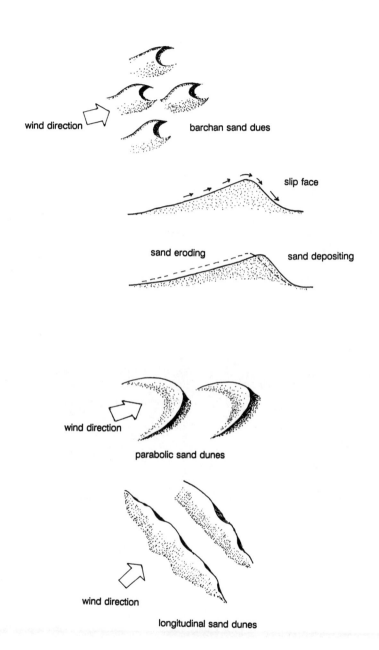

Fig. 5-32. Examples of several types of sand dunes and their development relative to wind direction.

155

Fig. 5-33. Classic barchan sand dune at the Bruneau Sand Dunes State Park, Idaho.

Fig. 5-34. Largest single sand dune in North America at Bruneau Sand Dunes State Park.

Fig. 5-35. Incipient pothole development in basaltic bedrock; note the asymmetry caused by current direction from left to right. Big Wood River, Idaho.

melting glaciers during the next few thousand years carried large sediment loads and were instrumental in developing the spectacular potholes and bedrock erosional features now found in the channel.

Most of the grinding tools responsible for pothole development are pebbles and cobbles of quartzite, plutonic and gneissic rocks transported more than 100 km from the mountains to the north. Almost all of these pebbles and cobbles are harder or more durable than the basaltic bedrock. Where a depression was sufficiently large to trap the pebbles or cobbles (Fig. 5-35), the current drove the loose rocks around in a circular pattern abrading the depression into a bowl-shaped feature or pothole. As the pothole enlarged, it became cylindrical in form with smooth polished walls.

As the potholes enlarged and expanded both horizontally and vertically, they coalesced with one another (Fig. 5-36). When a pothole captures the pebbles of a smaller adjacent pothole, growth is terminated in the smaller pothole and it is eventually consumed by the deeper pothole that acquired the pebbles (Figs. 5-37 and 5-38).

Fig. 5-36. Two potholes will ultimately enlarge and coalesce to form an elongate pothole. Big Wood River, Idaho.

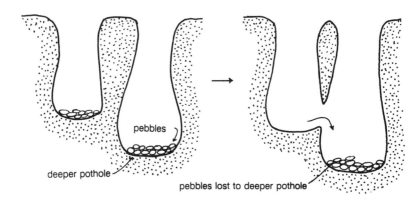

pebbles

deeper pothole

pebbles lost to deeper pothole

Fig. 5-37. Cross section of potholes shows how continuous enlargement or growth of adjacent potholes leads to merging. Where a deeper pothole breaches the bottom of a contiguous pothole, the pebbles are pirated from the shallower pothole and it is gradually cannibalized.

158

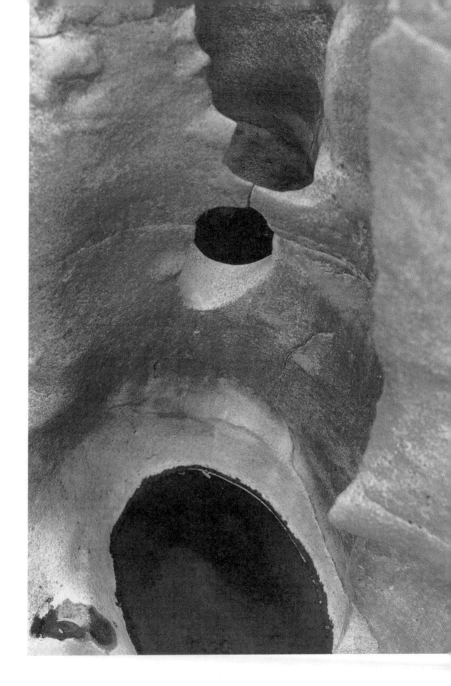

Fig. 5-38. Example of two merging potholes with pebbles captured by the deeper pothole. Without grinding tools (pebbles), the smaller pothole remains inactive and suspended on the wall of the still growing pothole. Big Wood River, Idaho.

Most of the basic erosional forms along the Big Wood River channel are created by numerous and diverse potholes in all stages of development and in a variety of shapes and sizes. Potholes range in size from 3 cm in diameter and 2 cm deep (Fig. 5-35) to as large as 10 m in diameter and more than 13 m deep.

The most common shape for a pothole is that of a teardrop, small at the top and large at the base. In fact the very largest potholes have a teardrop shape. Potholes also have pinch-and-swell forms and cylindrical forms.

All of the features within the channel are overprinted with a strong asymmetry (Figs. 5-39 and 5-40) caused by the abrasion of current-driven pebbles against the upstream side of the features. Consequently, the upstream side of the features tends to be smooth, convex and rounded (Fig. 5-41) and the downstream side tends to be concave with the leading edge of the feature pointing in the downstream direction.

Features Caused by Catastrophic Flooding

Two of the largest floods ever recorded on earth, the Missoula Flood and the Bonneville Flood, occurred in the northwest about 15,000 years ago. The Missoula Flood apparently occurred more than 40 times due to repeated damming of Lake Missoula by glacial movement. These catastrophic floods from glacial Lake Missoula, which swept across Montana, northern Idaho, Washington and Oregon, caused far more disturbance than did the Bonneville Flood. When the ice dam that contained impounded Lake Missoula failed, 500 cubic miles of water were suddenly released. The Bonneville Flood was significantly smaller, releasing a total of approximately 380 cubic miles of water. However, both floods left similar depositional and erosional features.

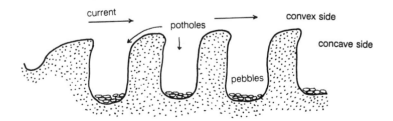

Fig. 5-39. Asymmetric forms in bedrock formed by current action. Forms are convex on the upstream side and concave on the downstream side.

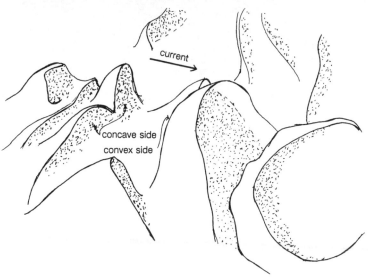

Fig. 5-40. Strong current from left to right gives distinct asymmetry to the forms. Forms are convex on upstream side and concave on downstream side. Big Wood River, Idaho.

current

concave side
convex side

Fig. 5-41. Asymmetric forms looking downstream; note how rounded or convex the forms are on their upstream sides. Big Wood River, Idaho.

Lake Bonneville was the precursor of the Great Salt Lake. The shorelines of this ancient lake can be seen on the higher slopes of the Wasatch Mountains more than 300 m above the present level of the Great Salt Lake (Fig. 5-42). Before the Catastrophic Bonneville flood, Lake Bonneville covered a large part of northwestern Utah.

Approximately 15,000 years ago Lake Bonneville, a late Pleistocene lake, suddenly discharged an immense volume of water to the north. This flood is thought to have been caused by capture of the Bear River which greatly increased the supply of water to the Bonneville Basin. The flood waters flowed over Red Rock Pass in southeastern Idaho and continued westward across the Snake River Plain generally following the path of the present Snake River. Although this enormous flood was first described in the literature by Gilbert in 1878, Harold Malde (1968) of the U.S. Geological Survey published the first detailed account of the effects of the flood on the Snake River Plain.

Fig. 5-42. Wave-cut bench along shoreline of pluvial Lake Bonneville, Northeast Utah.

Large rounded boulders of basalt characterize many deposits left by the flood along the Snake River Plain (Fig. 5-43). H.A. Powers, who recognized that these boulders were of catastrophic origin, and Malde applied the name of Melon Gravel to the boulder deposits (Malde and Powers, 1962). They were inspired to use this term after observing a road sign in 1955 that called the boulders "petrified watermelons."

The Melon Gravels deposited by the Bonneville flood average 1 m in diameter, but some well-rounded boulders range up to 3 m in diameter (Fig. 5-44). These boulders are composed almost entirely of basalt broken from nearby basalt flows. Only 10 to 15 km of transportation by the flood was needed to round the boulders. Melon gravels were dumped in unsorted deposits up to 100 m thick, 1.6 km wide and 2.4 km long (Fig. 5-46).

Fig. 5-43. Flood deposits are called melon gravels because the rounded cobbles and boulders look like melons. A farmer removed these "melons" from his field west of Twin Falls, Idaho.

Bretz (1923) first used the term "scabland" in reference to the eroded surface of basalt flows scoured by the Spokane Flood. This erosion was caused by glacial floods removing soil and rock from the surface. Scabland erosional features include coulees, dry falls (alcoves) and anastomosing channels distributed in such a manner as to cause a bizarre landscape (Figs. 5-47 and 5-48). The area affected by the flood can readily be established by the presence of scabland (Fig. 5-46). Gigantic dry falls and potholes range up to 40 meters in depth (Fig. 5-47). Minor features on the basalt bedrock exposed by the flood include polished and fluted surfaces that indicate the direction of the flow.

Fig. 5-44. Exceptionally large boulders of basalt, slightly rounded dropped by the flood waters on the north rim of the Snake River Canyon.

Fig. 5-45. Enormous boulder pile deposited by the flood. King Hill, Idaho.

Fig. 5-46. Looking north at scablands where all loose alluvial material was swept off the basalt bedrock. These scablands cannot be cultivated. The boundary between the cultivated land and the scabland also represents the boundary of the flood. North of Twin Falls, Idaho.

166

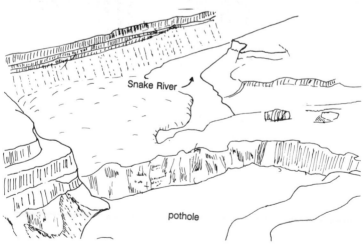

Fig. 5-47. Looking east at an enormous pothole enlarged by the Bonneville flood. It is situated at the foot of Shoshone Falls on the Snake River near Twin Falls.

167

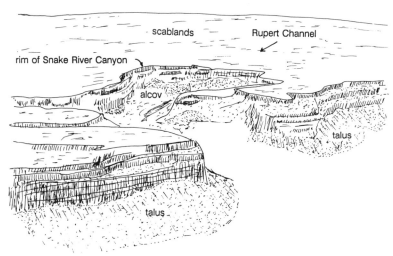

Fig. 5-48. Looking northeast at Blue Lakes Alcove along the north rim of the Snake River canyon. Note the scabland on the rim where the flood overflowed the canyon and swept across the Rupert channel, poured back into the canyon again and formed the alcove by cataract retreat.

168

PLACER DEPOSITS

A **placer deposit** is derived by mechanical concentration of heavy minerals from weathered rock. Weathering processes allow the mineral to be released from the source rock, and erosional agents such as wind and water transport and concentrate the minerals.

Stream Gradient

The upper portion of streams do not generally contain placer gold because the gradient is too steep; however, some coarse gold may be found in bedrock crevices. The best placers are formed where the gradient is just downstream from the steep gradient. Only very fine gold particles can be found where the gradient is low. Topographic maps prepared by the U.S. Geological Survey are excellent for determining the gradient of a stream.

Sources of Placer Minerals

Sources of placer minerals include veins, preexisting placer deposits, alluvial material with no placer concentration and sedimentary rocks such as conglomerate. If a placer deposit were derived from a lode deposit, the lode may have eroded away and no longer exist. One or more veins may have supplied the gold to a placer deposit. Preexisting stream placers are constantly reworked by running water, creating new placers which may or may not be richer than the original.

Chemical and Mechanical Weathering

Placer gold is derived from chemical and mechanical weathering of gold-bearing rock. Upon disintegration of the host rock, free gold is transported by running water. As rock fragments are transported downstream, they are broken into smaller particles and additional gold is released.

Water saturates all the sedimentary material in the stream bed. The turbulence and movement of water, particularly at periods of high runoff, allow the gold to work down to bedrock. Sedimentary materials tend to stratify according to density so that the minerals with the highest specific gravity (gold is about 7 times quartz and feldspar) quickly move down through the gravels until trapped on or close to bedrock. Minerals with a low specific gravity move faster and farther downstream than mineral particles with a high specific gravity.

Gold is generally separated from the primary deposit by chemical weathering. This is accomplished in one of three

169

ways: (1) the other minerals may be disintegrated and leached away leaving the gold to remain in the oxidized deposit; (2) the gold may be dissolved and removed from the deposit; and (3) the dissolved gold may be precipitated on gold particles (nuclei) as they are moved along in the alluvium; this may partly account for how nuggets are formed.

Minerals in Placer Deposits

The most common minerals to occur in placer deposits are garnet, diamond, chromite, quartz, muscovite, amphibole, pyroxene, tourmaline, rutile, barite, corundum, limonite, zircon, magnetite, ilmenite, cassiterite, wolframite, scheelite, cinnabar, pyrite, galena, sphalerite, arsenopyrite, native arsenic, native mercury, native silver, native copper, platinum, gold, molydenite, chalcopyrite, hematite, kyanite, topaz, spinel, allanite, epidote, sphene, tantalite-columbite and apatite. Gemstones such as diamonds, emeralds, topaz, garnets, sapphires and rubies also occur in placers. Placer minerals share several common characteristics: (1) they tend to be resistent to mechanical abrasion; (2) they tend to be resistent to chemical solution; and (3) they tend to be equidimensional in form.

Stratification and Sorting

Coarse gold is typically found with gravel-sized or larger sedimentary material; whereas, fine gold is found in silt, sand and pebble-sized material. Stratification and sorting are valuable indicators of the gold potential of a placer. Poorly sorted and stratified material such as glacial deposits generally are not gold bearing. There must be stratification or sorting of the material so that concentration can occur (Fig. 5-49).

Bedrock

The term **bedrock**, where used in connection with placer or alluvial mining, means a hard surface below which there is no valuable placer material. Bedrock can be represented by a great variety of rock types including granite (Fig. 5-50), basalt, sandstone (Fig. 5-51), schist, gneiss, quartzite, shale and conglomerate. Even a poorly consolidated clay or sand layer can be considered bedrock. Rocks that have steeply inclined bedding, fractures or cleavage, such as slates and schists, are effective at trapping gold. Rocks such as clay, volcanic tuffs and decomposed granite provide effective bedrock. If the bedrock is too smooth or flat, it will be ineffective because an irregular surface is necessary to trap the gold. In hard rocks,

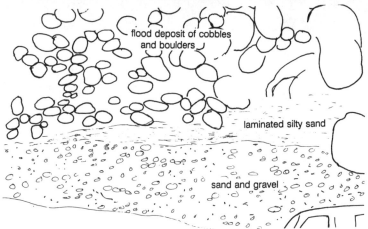

Fig. 5-49. Crude stratification in unconsolidated fluvial river deposits. The lowest layer consists primarily of silt, sand, pebbles and cobbles. Overlying that deposit is a thin laminated lens of silt and sand. And overlying the silt and sand is a thick deposit of cobbles and boulders, most likely deposited under flood conditions. The upper surface of the silt and sand mixture may act as a "false" bedrock and would be a good place to look for placer gold. Note that the pebbles, cobbles and boulders are rounded, as is characteristic of river gravels.

171

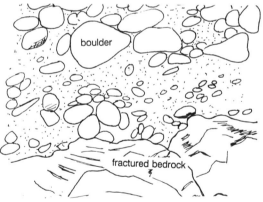

Fig. 5-50. A fluvial deposit of pebbles, cobbles and boulders overlying fractured granitic bedrock. This deposit has been mined for placer gold which occurs in the bedrock fractures. Miners were able to excavate into the bedrock about 0.5 m by wedging the fractured granite out with hand tools. Gold, which is approximately 20 times the specific gravity of water, works down into the crevices during the pounding and vibration by the river at flood stage. Near Atlanta, Idaho.

172

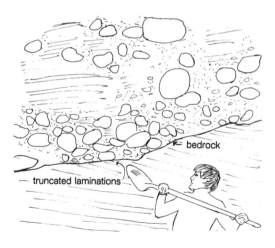

Fig. 5-51. Man is collecting sample at the interface between the laminated sandstone bedrock and the overlying gravel deposit to test for gold. This should represent the most likely place to find gold because the sandstone would act as bedrock and trap the gold. Lower Salmon River, Idaho.

173

the gold works into fractures and crevices; in softer materials such as decomposed granite it works into the bedrock. This is why the upper meter of bedrock should be evaluated. Normally bedrock will carry the greatest values in its depressions (Figs. 5-52, 5-53 and 5-54) and cracks. As a rule-of-thumb, approximately 90 percent of all gold in a placer deposit is concentrated within 0.5 m of bedrock—either above or below the gravel-bedrock interface.

False Bedrock

False bedrock can be a problem both during sampling and mining. Pay streaks situated on a well-defined layer of sand, gravel or clay above the bedrock are said to lie on a false bedrock (Figs. 5-49 and 5-50). False bedrock is simply a hard surface of dense material that is suspended above true bedrock. False bedrock can be very deceiving because it may consist of the same material as the true bedrock. Sampling to false bedrock will yield inaccurate sample results because the potentially higher-value gravels are not tested. By utilizing drilling and seismic reflection methods, it is sometimes possible to ascertain the depth of true bedrock. A seismic record can provide a continuous profile across the deposit and establish the profile of bedrock to determine the thickness of the gravels.

Gold Concentration Above Bedrock

Gold is rarely uniformly disseminated or distributed throughout a placer deposit. Fine gold in silt and sand deposits may come closest to approaching a disseminated deposit. If there is a continuing source of gold, the overlying alluvium may contain disseminated gold.

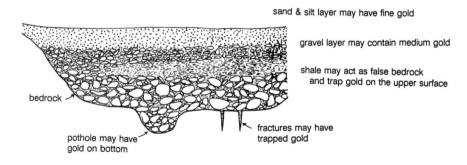

Fig. 5-52. Cross section of a stream showing typical sedimentary features and potential sites for high gold values.

174

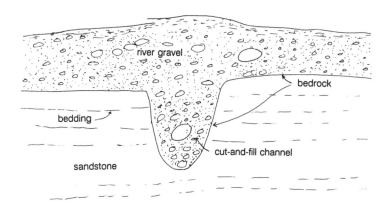

Fig. 5-53. Cut-and-fill channel of unconsolidated river deposit of sand, pebbles and cobbles cut in a thinly laminated sandstone. Gold or heavy minerals would most likely be concentrated on the sandstone bedrock in the lowest part of the channel.

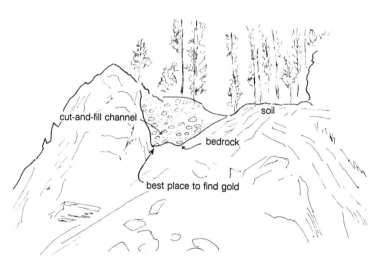

cut-and-fill channel

soil

bedrock

best place to find gold

Fig. 5-54. Abandoned river channel with lodge pole pine trees now growing over the channel deposits. Although difficult to locate, these paleochannels offer some of the few good opportunities left for small-scale placer gold mining.

176

Very fine gold, commonly referred to as **flood**, **flour**, **float**, or **skim** gold, does not concentrate easily by settling through sand and gravel. It may be transported hundreds of kilometers until it is deposited temporarily on the inside of meanders or in slack water. Flour gold is typically disseminated through great sequences of sediment. This type of deposit may appear, at first glance, to have high potential, but rarely meet expectations.

Gold is continuously concentrated on bedrock as a stream cuts downward. A stream performs like a sluice box, trapping the gold on the stream bed and allowing the water to carry the valueless, lighter materials downstream.

Paystreaks

Paystreaks are narrow, elongate-shaped bodies containing higher gold concentrations than surrounding materials. They may represent the original stream floor where gold was concentrated before base level was reached and the stream widened its valley (Figs. 5-53 and 5-54). The form is generally irregular and the enriched material may or may not have a different appearance than the surrounding material. Because the paystreak may represent the only commercial-grade gravels in a deposit, it is important for the prospector to ascertain if one exists in a deposit, and if so, what is its shape. A paystreak normally lies along bedrock, but may exist above bedrock, such as on a false bedrock. In fact it is not uncommon for two or more paystreaks to be superimposed on one another. Paystreaks tend to be sinuous in their lateral extent, most likely caused by deposition from a meandering stream channel. The richest paystreaks are produced by reworking preexisting gold-bearing gravels.

In narrow gulches and youthful streams, the paystreak is commonly in or near the notch of a V-shaped valley or gulch, on or near bedrock (Figs. 5-53 and 5-54). However, after a valley bottom has been widened from the migration of a stream, the paystreak may no longer occupy the lowest depression. It may be buried by later deposition of gravels and may have no relation to present streams. Paystreaks buried in wide valleys are the most difficult to establish. Their extremely erratic distribution is caused by shifting channels and migrating meanders.

Several methods are available to establish the position of a buried paystreak prior to mining. Among the methods available are profiling by trenching, overburden drilling and use of

177

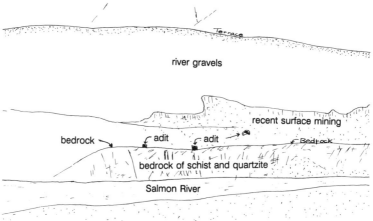

Fig. 5-55. Placer workings exposed along the main Salmon River north of Riggins, Idaho. Almost 100 m of unconsolidated river gravels overlie a quartzite bedrock. About the turn of the century, miners followed the paystreaks of gold through underground horizontal workings called adits. The workings were not supported with timbers even though the gravels are unconsolidated. These miners were well aware that the paystreaks of gold generally followed the low areas along bedrock, and that most gold is found near the bedrock-gravel interface. Using this mining method, they could extract most of the gold without mining the entire gravel deposit or disturbing the surface.

178

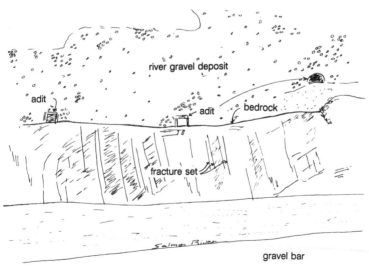

Fig. 5-56. Close-up placer workings shown in Fig. 5-55.

179

geophysical methods. If the overburden is thin and the bedrock magnetically low, a magnetometer survey may delineate the location of the magnetic black sands (magnetite) which may be associated with the gold. Under certain conditions, hammer seismic methods may be used for outlining the notches or channels in bedrock containing paystreaks. The early Chinese miners commonly drove adits along bedrock covered by thick sequences of gravel to follow the paystreaks (Figs. 5-55 and 5-56). By this mining method they were able to avoid processing or moving low-valued material. Another time proven method is to take a vertical channel sample or drill sample (Fig. 5-57).

Bars and Flood-Stage Deposits

Bars form in streams at places of relatively low velocity such as at or near the convex bank or the inside meander of a stream. Most of the gold and other material in a stream is moved during the flood stage so it is important to realize when prospecting stream gravels that the position of a paystreak is established during the flood stage.

Topography

Topography that indicates the best potential for commercial placer deposits is characterized by broad, terraced valleys and deeply weathered, rounded hills. Large placer deposits are not typically found in alpine topography characterized by high gradient streams and V-shaped valleys.

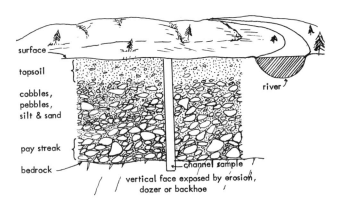

Fig. 5-57. Cross section through river gravel deposit showing how to take a vertical channel sample to test the value of the deposit.

180

Arid conditions do not favor the development of large placer deposits. The sporadic occurrence of cloud bursts sending large volumes of water down dry washes is not conducive to placer formation.

Size of Gold

The farther gold is transported from its source, the smaller the particle size. Near the source, gold particles tend to be rough and coarse, but with distance, they become smoother, more rounded and smaller. The term **color** generally refers to a small particle of gold that is visible to the naked eye. As a rule, coarse gold (nuggets) will remain on a 10-mesh screen; medium gold will pass through a 10-mesh screen; fine gold passes through a 20-mesh screen; and very fine gold passes through a 40-mesh screen.

Form of Gold

The external form, color, fineness and other features of placer gold visible to the naked eye are generally similar for any one source. In some cases it is possible to determine from which creek a specimen of placer gold originated. Vein gold normally has a high luster; whereas, placer gold is much more subdued. Manganese, iron oxides and iron humates may coat the gold giving it a black or deep brown coloration. Gold may be coated white or gray by calcium carbonate, colloidal silica or fine-grained clay. In addition to forming as equigranular colors, gold also commonly occurs as small scales or plates, nuggets, crystals, wires, tufts, filiform and dendritic forms.

Fineness of Placer Gold

Placer gold deposits range in **fineness** from 500 (50 % gold) to 999 (99.9 % gold). The other elements in placer gold include silver, copper and iron. As a general rule the fineness of placer gold will be higher than the gold in veins from which the metal originated. Also, the farther gold has been transported from the source and the smaller the particle size, the higher the fineness. For example, Snake River flour gold has an average fineness of 950. Gold in gossans and oxidized zones tends to have a higher fineness than gold in primary ores of veins.

Types of Placer Deposits

Residual placers are those where valuable minerals are concentrated at or near the source and the valueless lighter minerals are transported away by erosive forces such as wind, water and gravity (Fig. 5-58).

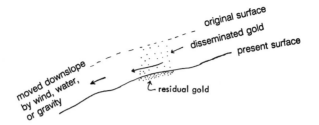

Fig. 5-58. Cross section of a residual placer. All of the gold that was once disseminated between the two surfaces is now concentrated on the lower surface.

Eluvial placers are generally found as a concentration of gold on a hillside slightly below a vein source. The lighter surface material may be removed by surface wash and wind. This type of placer generally occupies a small area. Eluvial placers are formed in weathered residual material, either overlying or in the vicinity of a primary gold deposit. These placers are formed, without stream action, on the slopes of hills where a lode deposit is deeply weathered. The heavier resistant minerals such as gold collect just below the outcrop and the lighter less resistant minerals are disintegrated and washed down slope or blown away by the wind. A large part of the lighter minerals are moved down slope by down hill creep. As a result of the removal of the valueless materials, the residual deposit is concentrated, although not as well or efficiently as could be done by running water. The gold in eluvial placers tends to be rough and irregular in form and have a fineness only slightly higher than that of the primary deposit.

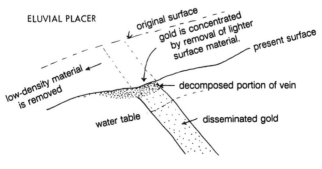

182

Gulch or stream placers are situated in or near active streams (Figs. 5-59 and 5-60). They are characterized by steep gradients and poorly sorted material with large boulders. Gold is coarse, concentrated on bedrock and occurs in crevices, potholes and in small paystreaks in gravel lenses or bars.

Bench placers are remnants of deposits left on a terrace or hillside where a stream once occupied a higher level than it does at the present. There may be several sets of benches (Figs. 5-61 and 5-62). Once gold is discovered in stream placers, you should investigate bench gravels above the stream for remnants of earlier stream gravels deposited before the stream cut to its present level. This type of deposit is often overlooked and represents a good place to prospect. Bench gravels may or may not lie on a terrace. In some cases bench or terrace gravels may be found by inspecting aerial photographs or topographic maps.

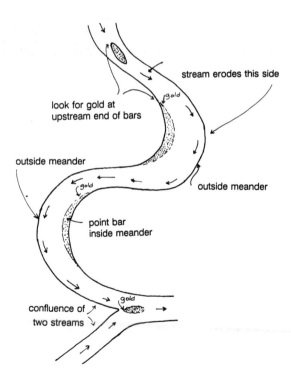

Fig. 5-59. Plan view of a stream showing where placer deposits are likely to occur.

183

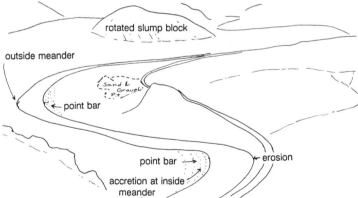

Fig. 5-60. Looking downstream along the Salmon River southwest of Challis, Idaho. The river is actively eroding material from the outside meanders where the current is strongest, and depositing in the slack water at the point bars along the inside meanders. The bedrock of the river is Precambrian quartzite. The hills surrounding the river are tuffaceous deposits of the 40-million-year old Challis volcanic rocks. Sand and gravel is mined from the river deposit in the west center of the photo. Also note the tilted or rotated slump block that slid off the hill in the upper left side of the photo.

184

Prospecting for bench gravels is particularly convenient along highways that follow streams. You may find road cuts that expose the contact between terrace or bench gravels and the underlying bedrock. Take your sample from gravels on the lowest point of the gravel-bedrock interface, using a shovel and canvas so as not to let small particles of gold escape.

Flood Gold deposits carry very fine gold particles which are transported during floods. The classic locality for flood-gold deposits is the Snake River in Idaho. In deposits of the Snake River the particles of gold are so fine that several hundred colors are only worth about one cent. Flood gold tends to be concentrated in the upper few centimeters of the gravel and at the upstream point of the bar. Although a few operations on the Snake River have been profitable for a short duration, there has been no sustained or significant production since such mining started in the 1860s. Flood gold gets its name because the fine gold can be transported long distances during flood conditions.

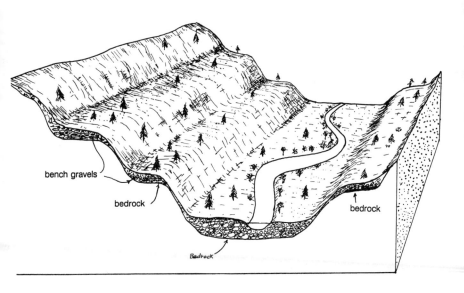

bench gravels

bedrock

bedrock

Bedrock

Fig. 5-61. Cross section of stream valley showing two levels of "bench gravels" left by the river as it cut deeper into the valley. These benches were cut by the river when it occupied a higher level.

185

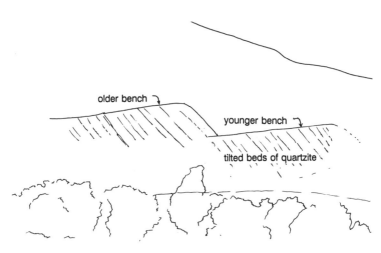

older bench

younger bench

tilted beds of quartzite

Fig. 5-62. Photo shows two distinct bedrock benches cut in quartzite by the Salmon river when it occupied a higher level in the canyon.

Eolian or desert placers form in desert regions where wind action removes low specific gravity material and enriches the surface in gold (Fig. 5-63). Concentration of gold is erratic and unpredictable because much debris moves off hillsides due to gravity, wind and sheet wash. Desert streams generally are violent in nature and move down gullies during infrequent rain storms. Such infrequent and sporadic water movements are ineffective in concentrating gold. The deposits tend to be small and rest on false bedrock such as caliche layers.

Glacial placers are formed by glaciers scraping off material from mountainous areas. Gold-bearing vein material is removed along with huge amounts of barren material. During this process, gold is mixed indiscriminantly with large amounts of glacial debris. However, glacial streams working through the glacial till have the capacity to sort the rock and mineral fragments according to size and specific gravity and concentrate any gold contained in the till (Fig. 5-64).

Beach placers are formed by wave and current action along the shorelines of lakes and oceans. Such placers tend to be best developed along rectilinear shorelines. Irregular and rocky shorelines are generally barren of such deposits. Magnetite and ilmenite represent most of the heavy minerals in beach placers. Clay beds and hard pan may provide a false bedrock upon which paystreaks can form. Most gold in beach placers is fine grained. Paystreaks are commonly several centimeters to a meter thick and less than 100 m wide. The paystreak normally follows the strandline of the beach in an erratic manner.

Tertiary gravels generally refer to the large gold-bearing gravels deposited about 50 million years ago. Many are lithified and buried by more recent gravels or volcanic rocks. Tertiary gravels have been mined effectively by drift and hydraulic mining.

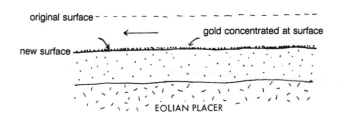

Fig. 5-63. An eolian placer is formed by wind removal of light sand and leaving gold concentrated at the surface.

Buried placers are formed where a placer deposit is buried by younger rock such as glacial debris, lava flows, eolian deposits, lacustrine deposits and marine deposits. If the base level of a stream is raised, gravels will probably be covered by renewed sedimentation of finer material, slumps and gravity flows of earth material. Volcanic flows and tuffs have covered many placer deposits. In many cases the old gravels are consolidated by cementation or compacted by the weight of overburden. Buried placers are commonly mined by underground "hardrock" methods. Miners followed the old stream channels with adits (horizontal underground workings).

A classic example of this type of deposit occurs near Idaho City, Idaho. In this area, a basalt flow covered stream placer deposits of Mores Creek (Fig. 5-65). Mores Creek has since cut down through the basalt, the buried gravels and the granite below. Now in several localities, the placer gravels of Mores Creek are sandwiched between the basalt flow and the underlying granite (Fig. 5-66). Early miners drove adits into the gravels and carried the gold-bearing gravels down to the creek to concentrate the gold (Fig. 5-67).

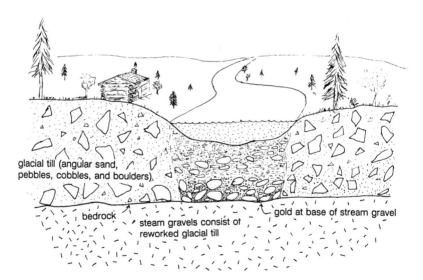

glacial till (angular sand, pebbles, cobbles, and boulders)

bedrock

steam gravels consist of reworked glacial till

gold at base of stream gravel

Fig. 5-64. Cross section of glacial till reworked by a stream. This type of deposit generally does not yield high gold values.

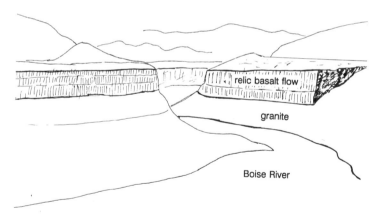

Fig. 5-65. The Boise River served as a channel for a basalt flow about 400,000 years ago. Since then the river has cut down completely through the basalt flow and another 10 m into the underlying granite. A remnant basalt wedge now lies on both sides of the river. The old river deposit, about 1 m thick, is sandwiched between the 100-million-year old granitic bedrock and the overlying basalt flow.

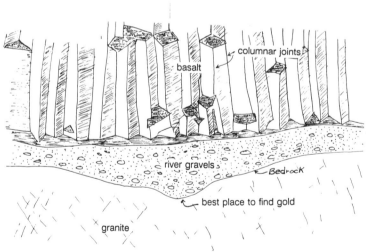

Fig. 5-66. A basalt flow covers paleo river gravels on granitic bedrock. Early Chinese miners extracted these placer deposits with underground workings.

190

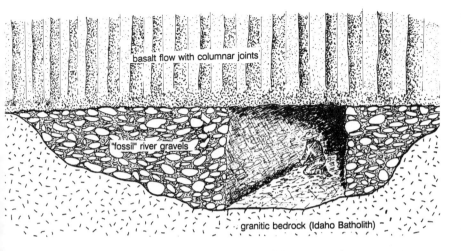

Fig. 5-67. The underground workings did not require support because the basalt flow acted as a ridged plate.

REFERENCES

Bretz, J.H., 1923, *The channeled scablands of the Columbia Plateau*: Journal of Geology, v. 31, p. 617-649.

Malde, H.E., 1968, *The catastrophic late Pleistocene Bonneville flood in the Snake River Plain, Idaho*: U.S. Geological Survey Professional Paper 596, 52 p.

Malde, H.E., and Powers, H.A., 1962, *Upper Cenozoic stratigraphy of the western Snake River Plain, Idaho*: Geological society of America Bulletin, v. 73, p. 1197-1220.

Maley, T.S., 1986, *Exploring Idaho Geology*: Mineral Land Publications, Boise, Idaho, 232 p.

Maley, T.S., and Oberlindacher, Peter, 1993, *Bedrock erosion in the lower Big Wood River channel, southcentral Idaho [abs.]*: Geological Society of America, Abstract with Programs, v. 25, no.5, p.113.

Maley, T.S., and Oberlindacher, Peter, 1994, *Rocks and potholes of the Big Wood River, south-central Idaho*: Information Circular 53, Idaho Geological Survey.

6 Volcanic Rocks

FEATURES OF BASALT

Types of Lava

Pahoehoe lava is characterized by thin sheets with low flow fronts; at the surface they are ropy, smooth or filamented (Fig. 6-1). This lava tends to have tongue-shaped lava toes and may grade into **aa lava** downstream as the viscosity increases.

Aa lava is blocky or platy and the fragments are covered with spines which gives it a very jagged surface; this lava tends to occur as piles of clinkers and slabs (Fig. 6-2).

Blocky lava has a surface of smooth-faced blocks.

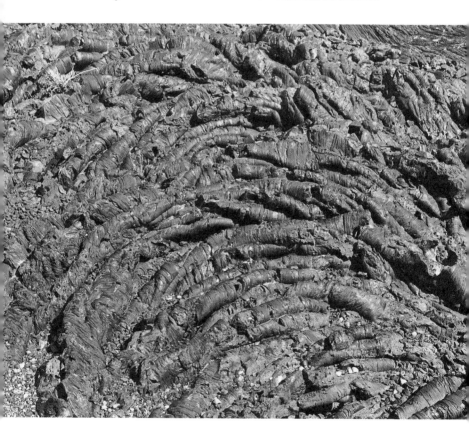

Fig. 6-1. Typical pahoehoe lava with smooth, ropy filamented surface. Craters of the Moon, Idaho.

Fig. 6-2. Smooth-surfaced pahoehoe basalt on the left and rough-surfaced aa lava on the right. Craters of the Moon National Monument.

Surface Features of Basalt

Shield volcanoes are formed by successive flows of basalt from a central vent. They pile up around the vent to form low, broad cones or shields. This type of cone is the most stable and the least susceptible to erosion.

Cinder cones are formed entirely of pyroclastic material, mostly of cinders. These cones consist of a succession of steeply inclined layers of reddened scoriaceous cinders around a central crater (Fig. 6-3). They are generally less than 350 m in height and are susceptible to erosion because there is nothing holding the mass together. Cinder cones have the steepest flanks of the three types of volcanic cones. They are also easily dismantled by lava extruded through the vent (Fig. 6-4).

Composite cones are formed by a succession of inter-layered basalt and cinders released from a volcano. These volcanic mountains are cone shaped and may be as much as 4,000 m high. Pyroclastic material is produced during periods of explosive activity and lava eruptions occur at times of quiescence.

193

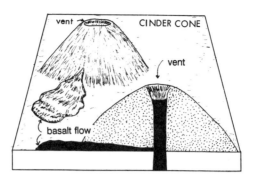

Fig. 6-3. Basaltic lava extruded through cinder cone.

Calderas are nearly circular basin-shaped depressions in the upper part of volcanoes. They are much larger than craters and are generally more than 10 km in diameter. There are two types: explosive calderas and collapse or subsidence calderas.

Spatter cones are relatively small cones formed by blobs of molten rock hurled out of volcanic vents. The hot, plastic blobs weld to the outer surface of the cone and quickly harden into rock (Figs. 6-5 and 6-6). **Hornitos** are small spatter cones.

Pressure ridges are elongate, low ridges of basalt caused by lateral compression from the downstream flow of lava (Fig. 6-7). They are aligned perpendicular to the flow direction and are convex in the downstream direction. Pressure ridges commonly form at the terminal edge of the flow (Fig. 6-8). They characteristically have a medial tension fracture running longitudinally along most of the ridge (Figs. 6-9 and 6-10). This crack may be caused by a final bulging of the lava below a brittle crust.

Pahoehoe lava tends to have more **vesicles** (air cavities) than does aa lava (Figs 6-11, 6-12 and 6-13). These vesicles form from degassing and are concentrated at the upper surface of the flow. **Pipe vesicles** may form pointing in the downstream direction. **Spiracles** are vesicles generally concentrated near the bottom of the flow from steam produced by hot lava running over wet ground surface.

Columnar joints, also referred to as contraction, shrinkage and tension joints, are generally well developed in basalt flows (Fig. 6-14). Joints normally form perpendicular to the cooling surface, both at the top and bottom of the flow (Fig. 6-15).

Lava tubes, which form only in pahoehoe flows, are important for the emplacement of lava because they are the

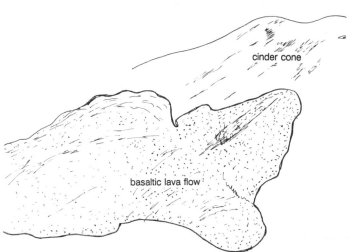

cinder cone

basaltic lava flow

Fig. 6-4. Cinder cone along the Great Rift. Following the eruption of cinders, basaltic lava flowed up through the vent and out through the flank of the cone. On the Great Rift near Craters of the Moon.

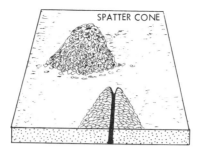

Fig. 6-5. Spatter cone built by small clots of lava piling up outside vent.

Fig. 6-6. Small spatter cone about 3 m high developed by clots of lava expelled from a vent and piled into a small cone. Craters of the Moon, Idaho.

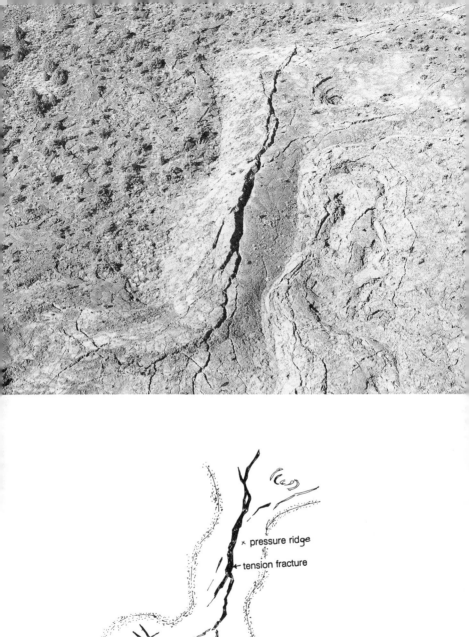

Fig. 6-7. A pressure ridge with a longitudinal tension fracture along its axis. Craters of the Moon, Idaho.

197

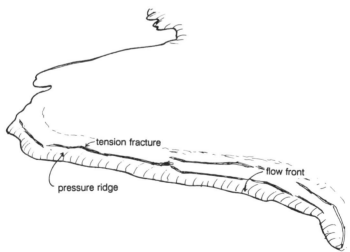

Fig. 6-8. A flow front of the 10,000-year-old Shoshone lava flow. The bulge at the margin is caused by lateral pressure from the flowing lava. The longitudinal tension fracture along the axis of the ridge was caused by the bulging of the brittle crust. Near Shoshone, Idaho.

subsurface passage ways that transport lava from a vent to the site of emplacement. Tubes originate from open flow channels that become roofed over with crusted or congealed lava. However a tube may also form in a massive flow (Figs. 6-16, 6-17 and 6-18). Lava tubes exist as a single tunnel or as complex networks of horizontally anastomosing tubes and may occupy up to five levels. Most tubes are 2 to 5 m across. Access to tubes is generally through collapsed sections.

Subaqueous Basalts

Many basalt flows, now exposed at the surface, were extruded into either lake or ocean bottoms. The deeper the basalt was emplaced below the water surface, the less vesicular and smaller the vesicles because of the high pressure. **Pillow lavas** have cross sections that give a pillow-shaped appearance. These pillows originate where a lava flow runs over a steep slope into water and the lava separates into discrete ellipsoids (Figs. 6-19 and 6-20). **Hyaloclastites** are the fragmental rock (breccia) in the matrix of pillows (Fig. 6-20). They are caused by lava quenched to glass then fragmented because of volume change; basaltic glass is altered to **palagonite** which is a very dark green, orange or brown.

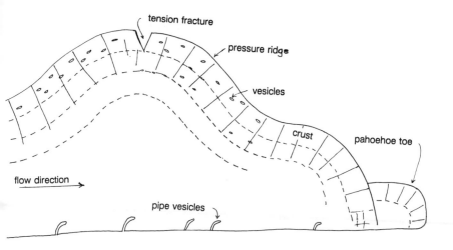

Fig. 6-9. Cross section of pressure ridge at the front of a basaltic lava flow.

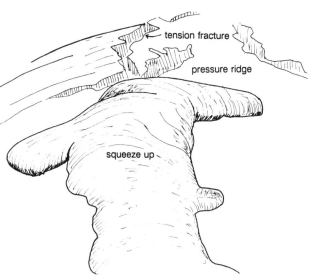

Fig. 6-10. A squeezeup formed by molten lava extruded through a tension fracture at the crest of a pressure ridge. Craters of the Moon, Idaho.

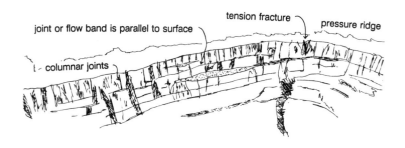

joint or flow band is parallel to surface

tension fracture

pressure ridge

columnar joints

Fig. 6-11. Cross section of pressure ridge of basalt with a set of flow bands parallel to the surface. For the most part, each plane of separation is a seam of aligned vesicles established during emplacement (flow lines). The uppermost layer is marked by a crude columnar jointing. This relationship indicates the columnar joints were developed during the cooling of the basalt but after the flow bands were formed. Northwest of Twin Falls, Idaho.

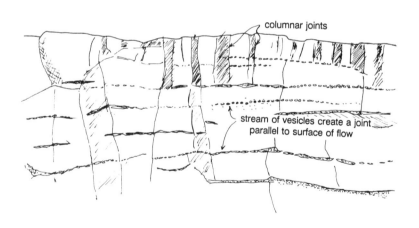

columnar joints

stream of vesicles create a joint
parallel to surface of flow

Fig. 6-12. Close-up of Fig. 6-11.

202

Fig. 6-13. Stretched vesicles typically occur in the 1- to 2-m-thick upper zone of a basalt flow. The vesicles in this low pressure zone are stretched while the lava continues to spread in a ductile state.

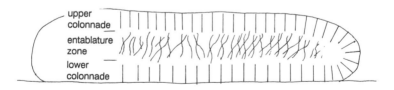

Fig. 6-14. Cross section of a lava flow showing an upper and lower colonnade of columnar joints and a central entablature zone. These joints, which form perpendicular to the cooling surface, are more widely spaced in the lower colonnade.

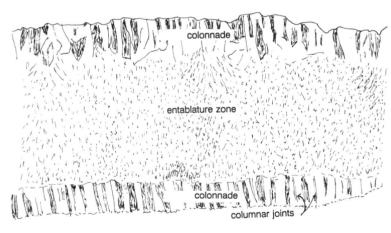

Fig. 6-15. Flow of basalt displaying two colonnades: one at the top and one at the base. The columnar joints of the upper and lower colonnades are perpendicular to the cooling surface. This particular flow has an entablature zone that is thick relative to the colonnades.

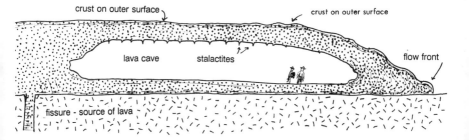

Fig. 6-16. Longitudinal view of lava flow with lava tube or cave.

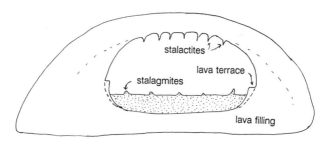

Fig. 6-17. Cross section of a lava tube within a basalt flow. Such a flow will consolidate and form a crust on the outside and the fluid lava in the core of the flow will continue to withdraw leaving a shell of lava.

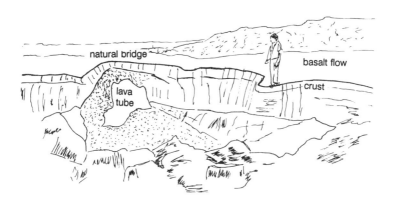

Fig. 6-18. Cross section of a lava tube formed by a collapse of the roof on both ends leaving an arch or natural bridge. Near Atomic City, Idaho.

Spheroidal Weathering

Spheroidal weathering is fairly common in some basalt flows. Those flows that have a large feldspar component and relatively coarse crystallinity are the most likely candidates (Figs. 6-23, 6-24, 6-25 and 6-26).

The Great Rift System

The Great Rift system, consisting of a series of north-northwest-trending fractures, extends from the northern margin of the eastern Snake River Plain, southward to the Snake River (Fig. 6-27). The Great Rift is a 100-km-long and 2- to 8-km-wide belt of shield volcanoes, cinder cones, lava flows and fissures (Fig. 6-28). Three lava fields are aligned along the rift: Craters of the Moon, Wapi and Kings Bowl. Craters of the Moon lava field covers an area of 1660 square km and is the largest Holocene (less than 10,000 years old) lava field in the conterminous United States.

Lava fields of recent basalt can be found at five different locations along the Great Rift: (1) Cerro Grande and other flows near Big Southern Butte; (2) Hells Half Acre lava field near Blackfoot; (3) Wapi lava field; (4) Craters of the Moon lava field; and (5) King's Bowl lava field. The younger flows lack vegetation so that they clearly stand out on aerial photographs.

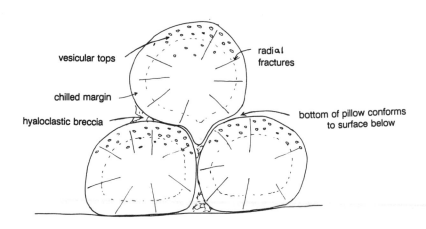

Fig. 6-19. Cross section of pillow lava showing common features.

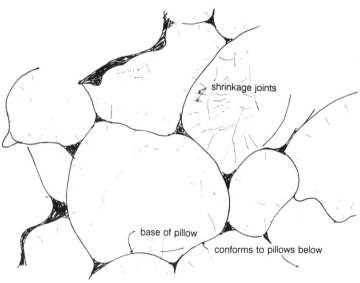

shrinkage joints

base of pillow

conforms to pillows below

Fig. 6-20. Pillow basalt near Grangeville, Idaho in Columbia River Basalt. Note how each pillow conforms to the surface below.

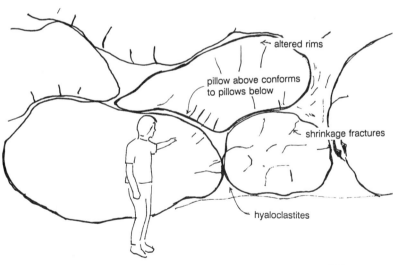

altered rims

pillow above conforms
to pillows below

shrinkage fractures

hyaloclastites

Fig. 6-21. Pillow basalt in Columbia River Basalt near Grangeville, Idaho. Note the bottoms of pillows conform to the surface below. These pillows have altered rims and hyaloclastites in the matrix.

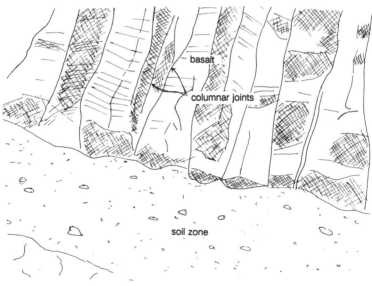

Fig. 6-22. A well developed soil zone is formed between two flows of Columbia River Basalt.

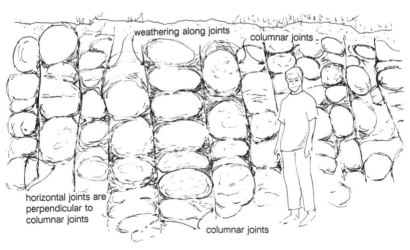

weathering along joints

columnar joints

horizontal joints are
perpendicular to
columnar joints

columnar joints

Fig. 6-23. Exceptional example of spheroidal weathering in basalt flow. Weathering was facilitated by the columnar joints and horizontal joints that provide access to the water. Near Grangeville, Idaho.

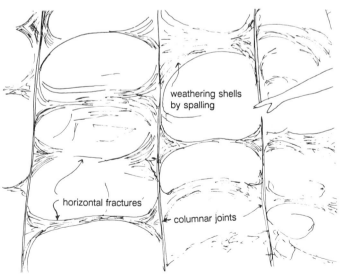

weathering shells
by spalling

horizontal fractures

columnar joints

Fig. 6-24. Close-up of Fig. 6-23.

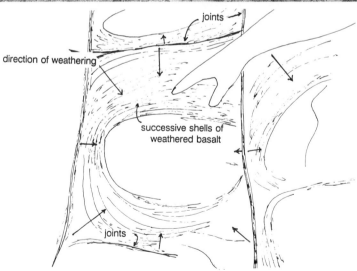

Fig. 6-25. Another close-up of Fig. 6-23. In this photo you can easily see how each successive shell becomes more spheroidal. Near Grangeville, Idaho.

213

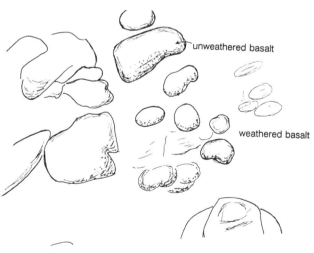

Fig. 6-26. Outcrop of highly weathered basalt. This spheroidal weathering has left only a few ellipsoidal relics of unweathered basalt. Note the weathered shells surrounding the relics of unweathered basalt. Near Idaho City, Idaho.

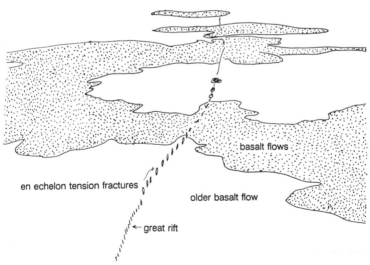

Fig. 6-27. Looking along the northeast-trending Great Rift. This rift can be followed about 100 km across the central Snake River Plain. Basalt was extruded from the fracture and flowed both east and west creating a mirror-image effect.

215

Fig. 6-28. A segment of the Great Rift with cinder cone developed over the rift. Both pyroclastic material and basalt escaped through the fissure.

Craters of the Moon National Monument

Craters of the Moon is a veritable outdoor museum of volcanic features (Fig. 6-29). Features such as crater wall fragments, cinder cones, mini-shield volcanoes, lava bombs, spatter cones (Fig. 6-30), lava tubes, tree molds and rifts are among the best examples, both in variety and accessibility, that you can observe anywhere in North America. The total field includes more than 60 lava flows and 25 cinder cones.

Ages of the Lava Fields. The lava flows in the Great Rift area were dated with charcoal from tree molds and charcoal beneath lava flows. This charcoal gave radio-carbon ages of 2,222 years old, plus or minus 60 years, for the King's Bowl lava field. Paleomagnetic measurements have also been used to correlate lava flows and determine the volcanic history in the area. The flows were extruded during eight eruptive periods beginning about 15,000 years ago until the last

216

Fig. 6-29. The triple twist tree which grew out of a fracture in the North Crater Flow has 1350 growth rings. Such evidence can serve as a minimum age for a flow in the absence of radiometric dates. Approximate relative ages of flows can, in areas like Craters of the Moon, be determined by the amount of surface weathering as indicated by coloration, vegetation growth and soil development.

eruption about 2100 years ago. Each eruptive period lasted several hundred years with a separation between eruptions of several thousand years.

Lava Flows. At Craters of the Moon, you can see excellent examples of the three types of basaltic lavas: blocky, aa and pahoehoe (Fig. 6-2). Pahoehoe flows are characterized by thin sheets with low flow fronts. At the surface pahoehoe flows are ropy, smooth or filamented (Fig. 6-2). The upper surface of the fresh pahoehoe is made of a dense to vesicular glass. This surface is commonly a greenish blue or an iridescent blue. The Blue Dragon flows, which cover most of the surface east of the Great Rift, have an unusual iridescent-blue color. In a single flow, one lava type can change to another; for example, pahoehoe lava can change into aa lava when it cools, becomes more viscous and loses gas.

For the most part, the flows in the Craters of the Moon consist of pahoehoe-type lava and were fed through a system of lava tubes. In many locations, the roofs of lava tubes have collapsed, leaving a mass of broken lava and rubble on the floor of the tube. Many such collapsed areas in the Broken Top and Blue Dragon flows provide "skylights" and entrances to lava tunnels for the numerous visitors to the monument.

The older flows in Craters of the Moon lava field can be determined by the nature of the surface. Younger flows have unweathered glassy crusts and a surface that is typically blue. Older lavas tend to be covered by wind-blown deposits and their surface is light colored from weathering and oxidation.

Cinder Cones. More than 25 cinder cones are aligned along a 27-km-long segment of the Great Rift (Figs. 6-28 and 6-30). Cinders are very vesicular clots of lava blown out of a vent in the earth's surface. Cinder cones are developed by the accumulation of cinder and ash in cone-shaped hills. Many cinder cones at the monument are a composite of two or more cones

Fig. 6-30. Three spatter cones and a cinder cone aligned along the Great Rift, Craters of the Moon.

with overlapping craters and flanks. These cones consist of agglutinated and nonagglutinated ash layers typically inter-layered with a few thin lava flows. Some cones are asymmetric or elongate reflecting the wind direction at the time of the eruption. Many cinder cones have been breached on one or more flanks as a result of erosion by lava (Figs. 6-3 and 6-4).

Volcanic Bombs and Spatter Cones. Volcanic bombs are blown from vents as clots of fluid lava. As they move through the air while still hot and plastic, they deform into aerodynam-ically shaped projectiles of lava. Spatter cones are relatively small cones formed by blobs of molten rock hurled out of volcanic vents. The hot, plastic blobs weld to the outer surface of the cone and quickly harden into rock (Figs. 6-5, 6-6 and 6-30).

FEATURES OF RHYOLITIC LAVAS

Shape and Size of Flows

Because rhyolite flows are generally much more viscous than flows of basaltic composition, they tend to be thick, bulbous and have a surface characterized by spines and upturned slabs. An individual flow of rhyolite may cover only a few square kilometers; whereas, ash-flow tuffs, also of rhyolitic composition, flow very rapidly and a single flow may cover hundreds of square kilometers.

Fractures

Rhyolite flows may have shrinkage or cooling fractures perpendicular to the flow surface similar to columnar jointing in basalt flows. In some cases there may be two sets of columnar joints: one related to the bottom surface and one to the top surface. Columnar jointing in rhyolite is rarely as well developed as it is in basalt flows. Extension fractures on the surface may be used to ascertain the flow direction which would be perpendicular to the fractures. Platy jointing or subhorizontal sheeting joints may be formed in the central zone and have a spacing ranging from 1 cm to more than 1 m (Fig. 6-31). Platy joints may or may not be related to flow banding or flow foliation.

Flow Bands or Flow Foliation

The best evidence of the last flow direction just before final emplacement and congealing of the rhyolite may be flow banding with its associated folds (Fig. 6-32). When mapping rhyolite flows, pay particular attention to the location, shape and distribution of folded flow bands or layers. Fold axes are normally perpendicular to flow direction, so the attitude of fold axes and axial planes (Fig. 6-33) should be recorded. Primary flow foliation and lineation where formed by aligned crystals, inclusions, spherulites, vesicles, schlieren (mineral segregations) should be described in terms of concentration, distribution and relationship to other features and rock types. Striations on a flow surface may indicate the last flow direction.

Zoning

A rhyolite flow may or may not have zoning; and if present, such zoning may range from crude to well developed (Fig. 6-34). A **basal** zone might be characterized by a basal vitrophyre layer and a basal breccia. Breccia, formed by the breaking up

Fig. 6-31. Folded sheeting joints in rhyolite showing surface lineations on the sheets. Note the basal breccia on the left side surrounding the sheeted lobe. Near Jordan, Oregon.

of the crust at the flow front, is overridden by the flow and becomes a basal breccia. A basal breccia may occur at the flow front, between the lobes, at the base of the flow and incorporated in the flow (Fig. 6-31). Breccia should be described in terms of size, shape, types of fragments and color. The distribution, thickness and character of vitrophyre and pumicious layers and fragments should be described as should the size, shape and type of devitrification spherulites which tend to be common in vitrophyre. The **central** zone might be characterized by lithoidal lava with well-developed shrinkage fractures and perhaps layers of sheeting joints. The **upper** zone might

Spherulites are speroidal masses of **lithoidal** lava that may have concentric or radial internal structure. They form from **devitrification** of glassy volcanic rock. Devitrification is the conversion from glassy to crystalline rock. Hollow spherulites may be later filled with chalcedony to form **thunder eggs**.

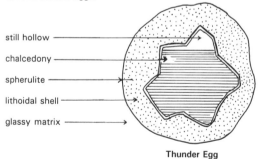

still hollow

chalcedony

spherulite

lithoidal shell

glassy matrix

Thunder Egg

Fig. 6-32. Typical folding of rhyolitic flow bands.

Fig. 6-33. Small refolded, tight isoclinal fold in rhyolite; note the well defined flow foliation surrounding the fold. Such folds are useful in determining movement direction of the lava. Snake river Canyon near Twin Falls, Idaho.

be characterized by abundant gas cavities (Figs. 6-35 and 6-36) which may be filled or partly filled with opal or jasper deposits. Record the distribution, size and shape of gas cavities. Flow tops and bottoms tend to be vesicular and vesicularity tends to increase upwards. Gas cavities may be flattened, distorted or stretched.

Obsidian

Obsidian forms when magma of rhyolitic composition cools so fast that crystallization of minerals is not possible (Figs. 6-37 and 6-38). Volcanic glass, essentially a frozen liquid, is a lustrous, glassy-black or reddish-black rock with a conchoidal fracture giving it very sharp edges.

Jasper and Opal Deposits

Jasper and opal deposits are commonly found associated with rhyolitic flow rocks. Silica, in the form of opal, chalcedony or jasper, is generally derived by leaching from the still hot rhyolite flow by hot water and precipitated in cool cavities. These silica deposits are typically found in openings left in the flow such as around breccia fragments, shrinkage fractures, spherulite shrinkage cavities and gas cavities (Fig. 6-39). Jasper and opal are most commonly found at the outer margins of a flow.

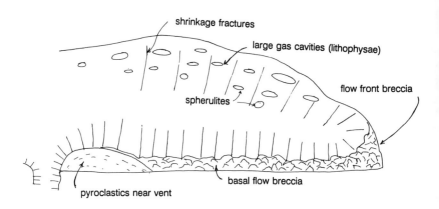

Fig. 6-34. Cross section of a rhyolite flow showing typical features.

Fig. 6-35. Vesicles in rhyolite. Geologist is searching for topaz crystals in the cavities. Topaz Mountain, Utah.

Fig. 6-36. Large vesicle or air cavity in rhyolite more than 1 m in diameter. Topaz Mountain, Utah.

225

Fig. 6-37. Flow bands in obsidian defined by thin layers of small vesicles and crystals.

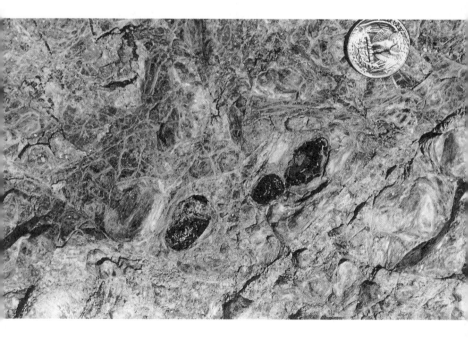

Fig. 6-38. Ellipsoidal apache tears of black obsidian, 1 to 2 cm in diameter, occurring in perlite. Perlite is a glassy igneous rock with the composition of rhyolite and spheruloidal cracks formed during cooling.

Fig. 6-39. Air cavities in rhyolite partially filled with precious opal. Near Spencer, Idaho.

FEATURES OF PYROCLASTIC DEPOSITS

Pyroclastic deposits are formed by explosion out of a central vent. They include: (1) airfall deposits are blown out of the vent and settle to the ground; (2) pyroclastic surge deposits or base surge deposits; they are pushed out by strong explosions and are transported outward at high velocities; and (3) pyroclastic or ash flow deposits are formed of hot debris and flow downslope.

Airfall Deposits

Airfall deposits are characterized by an even blanket-type covering of topography. These deposits drape over the surface with a uniform thickness. Near the source they are poorly sorted, but with distance, sorting improves. A succession of eruptions accompanied by ash fall activity will form a stratified deposit. Some welding and columnar jointing may occur, especially near the vent where the deposit is thick and initially hot.

These deposits are extremely susceptible to wind and water erosion. They are quickly removed from high areas and redeposited in the valleys and depressions. They may even be stratified as water-laid tuff.

Pyroclastic deposits may also be deposited by **lahars** (Fig. 6-40) or water saturated **volcanic debris flows**. Lahars commonly originate on the slopes of the source volcanos where the pyroclastic material is thick. Precipitation may saturate the deposit with water and cause it to be unstable. Once this saturated pyroclastic material starts moving downslope following the low topography, it has the energy to pick up much material in its path including trees and other forms of life, and even large boulders. Such deposits are poorly sorted and lack welding and compaction.

Pyroclastic Surge or Base Surge Deposits

Pyroclastic surge deposits originate from volcanoes and move laterally downslope as a hot, low concentration of gas-solid mixture. The surge blasts can be so violent that they knock trees down in their path. These deposits may blanket the topography but tend to be better developed in valleys. The direction of movement may be ascertained by studying the sedimentary flow forms such as cross bedding, dunes and antidunes, and truncation structures. Planar laminations are particularly well developed in these deposits (Fig. 6-41).

Ash-Flow Tuffs (Ignimbrites)

Ash-flow tuffs or ignimbrites are formed from volcanic eruptions of density currents with a mixture of hot gases, small globs of lava, hot ash and pumice fragments, all held in suspension. Once this "glowing cloud" of material flows out of a volcano, it moves downslope following the depressions in the topography at speeds of more than 100 km/hr. These deposits are widespread throughout the western United States and may even erupt and flow underwater. Where there is a succession of flows, the first few flows will fill in the topography. Successive flows will tend to pile up like a layer cake with constant thickness and a large horizontal distribution. They are particularly noticeable as rim formers. The composition ranges from rhyolite to dacite. Ash flow tuffs are characterized by poor

Fig. 6-40. Mudflow deposit filled with numerous, well rounded boulders of quartzite. Near Challis, Idaho.

Fig. 6-41. Laminated base surge deposit of ash.

sorting, lack of bedding and a variety of gradational features. Cooling units vary in cooling rate, amount of compaction, fusing or welding thickness. Fused layers may be distinct or grade into nonfused layers.

Zoning

Ash-flow tuffs tend to be vertically zoned (Figs. 6-42 and 6-43). These zones are based on differing (1) compaction, (2)

welding, (3) vapor caused alteration and (4) crystallization. A typical zone sequence from bottom to top includes the following:

1. **Basal bedded ash** is characterized by graded beds, strong lamination and cross beds; such ash may be formed by base surge deposition.

2. **Welded basal vitrophyre** is characterized by black glassy appearance and spherulites.

3. **Massive central zone** is characterized by dense lithoidal rock with few vesicles; this zone commonly has subhorizontal sheeting joints (Fig. 6-44).

4. **Upper zone** is characterized by subparallel sheeting joints commonly deformed into folds (Fig. 6-45); this zone commonly contains stretched vesicles, flow marks and some vitrophyre. The axes of primary flow folds are perpendicular to the lineations.

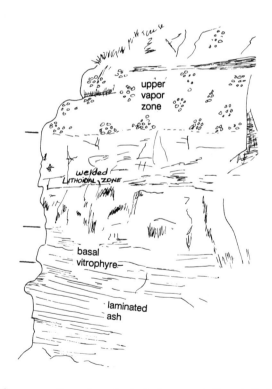

Fig. 6-42. Cross section of an ignimbrite or ash-flow tuff showing typical zoning.

Compaction

From a combination of the weight of the deposit and welding of the material as it settles, it forms a dense glassy to stoney, lithoidal rock. The welded portion may be characterized by a **eutaxitic** fabric (Fig. 6-46). This fabric forms from glass shards and pumice fragments compressed and flattened into dark lenticular features by the weight of the overlying mass. In some cases these compaction features may give the deposit a foliated appearance.

Welded Zones

Most deposits grade vertically from densely welded to only slightly welded near the surface. Portions may be so glassy that they break with a conchoidal fracture. The welded portion is characterized by columnar jointing and tends to form

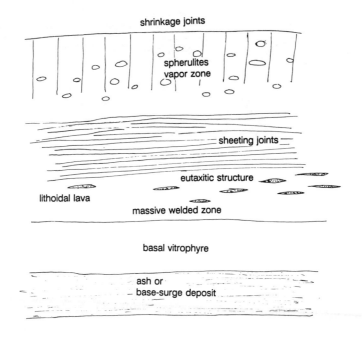

Fig. 6-43. An ash flow tuff with typical zoning including a base surge deposit, a basal vitrophyre, and a welded tuff unit with an upper vapor zone. Near Oakley, Idaho.

233

Fig. 6-44. Weathered outcrop of welded tuff with well-developed subhorizontal sheeting joints; also note the steeply dipping joint set. Flow banding intersects sheeting at an oblique angle at this outcrop. Gooding City of Rocks, Idaho.

vertical cliffs. Pumice fragments are compressed into lenticular **fiamme** of glassy obsidian. These glassy masses tend to be dark and contrast with the rest of the rock.

Crystallization and Vapor Effects

Crystallization and vapor phase alteration most commonly affects the upper portion of the welded and unwelded zones. This zone is characterized by large spherulites and lithophysae which generally increase in size and density upwards. Vapor phase crystallization may cause these vesicles to be coated with small crystals of sanadine, cristabolite or tridymite. Devitrification is most active in this zone.

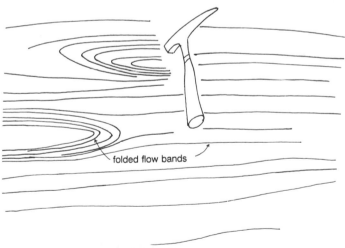

folded flow bands

Fig. 6-45. Small passive flow, intrafolial folds, defined by primary flow foliation in ash flow tuff. Folds were formed after emplacement and fusing but while the flow was still hot and ductile. The trends of the fold axes are useful for determining flow direction. South of Oakley, Idaho.

Fig. 6-46. Eutaxitic structure consisting of lenticular, light streaks in a welded tuff. These lens-shaped features were stretched while the tuff was still viscous.

7 Intrusive Rocks

Batholiths
Batholiths are large composite bodies of igneous rock ranging in composition from diorite to granite. They are predominantly medium grained, have hypidiomorphic granular texture, are megacryst bearing and are locally foliated, particularly at the margins. They are characterized by plagioclase, orthoclase feldspar, quartz, biotite, muscovite, hornblende and accessory minerals such as epidote, sphene, apatite, zircon, magnetite, and ilmenite.

Igneous Derived Granites (I-Type)
Igneous-derived granites originate from partial melting of the younger eugeosynclinal sedimentary and volcanic rocks, including the mantle and oceanic crust. They are characterized by low Sr87/Sr86 ratios with values generally less than 0.708. The most common rock type is diorite and distinctive minerals include hornblende, biotite, magnetite and sphene. Xenoliths are igneous and they are located near and associated with island arcs, continental margins and subduction zones.

Sedimentary Derived Granites (S-Type)
Granitic rocks derived from sedimentary rock originate from partial melting of old crystalline basement and the deep continental crust. They are characterized by high Sr87/Sr86 ratios with values generally more than 0.710. The most common rock type is granite and distinctive minerals include both muscovite and biotite (two-mica granites), cordierite, monazite and garnet. Xenoliths are typically metasedimentary rock. They tend to be associated with continental collision areas and overthrust terrains.

PLUTONS

A pluton is any large igneous intrusion that is not a sheet intrusion. A **Stock** is a pluton less than 100 sq. km in outcrop area; and a batholith is a pluton with more than 100 sq. km exposed.

Internal Zonation of Plutons
Plutons tend to have a crude internal zonation. They are typically mafic at the borders and grade to more silicic rocks towards the center. Early formed high temperature minerals

such as calcic plagioclase, magnesium-rich pyroxenes and amphiboles accumulate near the walls. The primary granodiorite magma may fractionate from diorite or tonalite at the margins to granite at the center.

Granitization

Granitization is any process by which a rock is made more like a granite. This process could be accomplished by diffusion of fluids through preexisting solid rock by ionic replacement along grain boundaries. Granitization could only account for a very small part of the earth's granitic rock and certainly could not produce an entire batholith. Granitization is a metamorphic process that occurs in the infrastructure at very high temperatures and pressures where both the texture and the mineral composition are changed.

Migmatization

Migmatites are mixtures of igneous or igneous appearing rock and metamorphic rock. They can originate by metamorphic segregation, intrusion and metasomatism. Some common examples include (1) lit-par-lit (Fig. 7-1), (2) porphyroblastic, (3) veins and nodes (Figs. 7-2 and 7-3), (4) ptygmatically folded (Fig. 8-1), (5) agmatite (Fig. 7-4), and (6) folded and swirled.

Emplacement of Granitic Plutons

Granitic plutons are emplaced by one or more methods including (1)forcible emplacement, (2) stoping, and (3) zone melting.

Forcible emplacement is more effective at great depths where the pressure and temperature are high and the rocks are ductile. As the relatively less dense granitic magma moves upward through the crust, it pushes the overlying rock aside and the country rock fills in behind. Evidence of forcible emplacement includes the following: (1) regional foliation (rock layers, bedding, schistosity, etc.) wrapped concordantly around the pluton; (2) foliation in the border zone of the pluton concordant to the foliation of the country rock; and (3) boudinage and recumbent folds.

Stoping occurs where the upward force of the magma fractures the overlying rock. Magma is injected into the fractures and then engulfs large blocks of country rock. Assimilation of the country rock changes the composition of the magma and adds to the volume. Stoping is more effective in brittle rocks at shallow depths. Evidence for stoping includes the following: (1) structures in the country rock discordant to

Fig. 7-1. Lit-par-lit injection gneiss. An igneous intrusion has penetrated along the cleavage planes in the quartzite leaving numerous thin parallel sheets of granitic rock between layers of quartzite. Because this is a combination of igneous and sedimentary rock, it is a migmatite. Note the alignment of biotite grains (secondary) in the granitic material parallel to the cleavage of the quartzite. Most likely, the cleavage in the quartzite was developed simultaneously with the intrusion of the granite and the development of secondary foliation in the granite. These features are typical of mesozonal levels of intrusion. Near Oakley, Idaho.

Fig. 7-2. Migmatite with veins of granitic material in conformable contact with biotite quartzite. Near Elk City, Idaho.

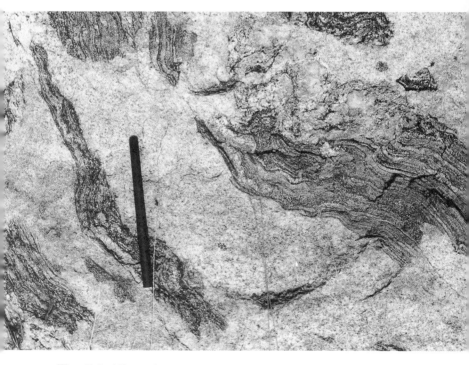

Fig. 7-3. Migmatite comprised of schistose xenoliths in granitic material.

Fig. 7-4. Intrusive breccia with fragments of country rock consisting of dark biotite schist in a granitic matrix. Each fragment of biotite schist has a mantle about 1 cm thick derived from the granitic intrusion.

contacts of the pluton; (2) xenoliths of country rock in the pluton; and (3) stoping along ring fractures.

Zone melting occurs where magma rises by melting and assimilating roof rocks.

Magmatic Emplacement Versus Granitization

Magmatic emplacement is characterized by (1) sharp contacts (Fig. 7-5), (2) rotated inclusions (Fig. 7-6), (3) chilled margins, (4) dilation offsets (Fig. 7-7), and (5) absence of relict bedding in sedimentary gneisses (paragneiss). Granitization or metasomatism is characterized by (1) gradational contacts (Fig. 7-8), (2) irregular boundaries (Fig. 7-9), (3) absence of chilled margins, (4) relict bedding in paragneiss, and (5) relics or **skiaths** across boundaries (Fig. 7-10).

Infrastructure or Catazonal Emplacement

The emplacement of a pluton in the infrastructure or catazone is characterized by the following features: (1) high

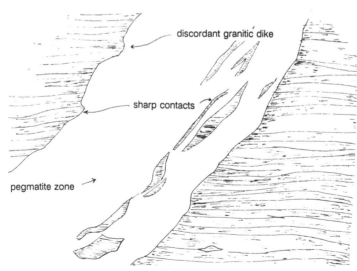

Fig. 7-5. Discordant granitic dike in strongly foliated metamorphic rocks. The sharp contacts indicate an intrusive igneous origin. Near Elk City, Idaho.

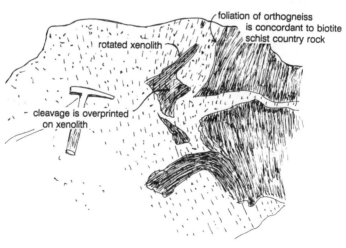

foliation of orthogneiss
is concordant to biotite
schist country rock

rotated xenolith

cleavage is overprinted
on xenolith

Fig. 7-6. Intrusive contact of granitic rock and foliated country rock. Note that the inclusions ripped off the adjacent schist are rotated. This is evidence that the granitic rock is intrusive rather than formed by metasomatism. Near Shoup, Idaho.

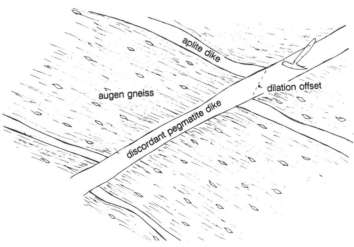

Fig. 7-7. Dilation offset of two aplite dikes by a younger pegmatite dike. This is evidence for an intrusive igneous origin for the pegmatite dike. Near Shoup, Idaho.

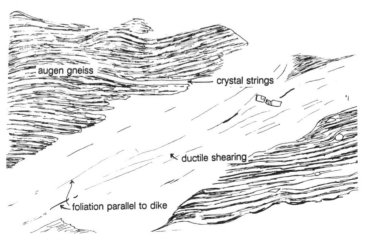

augen gneiss

crystal strings

ductile shearing

foliation parallel to dike

Fig. 7-8. Discordant granitic dike in strongly foliated augen gneiss near Elk City. The foliation in the central part of the dike is approximately parallel to the dike and possibly caused by ductile shearing within the dike. At the contact, the dike material appears to bleed along light layers of the augen gneiss.

245

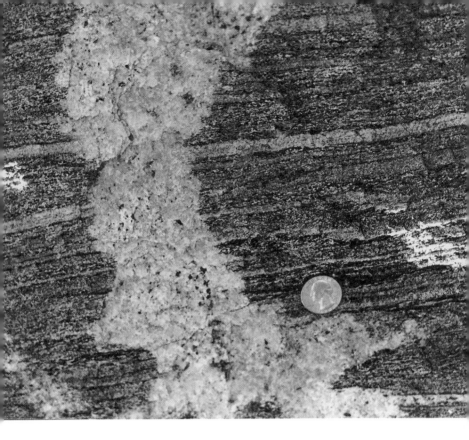

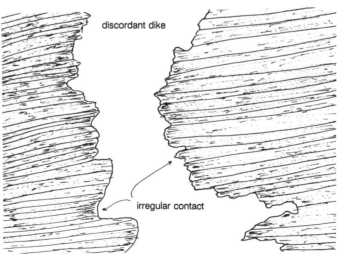

discordant dike

irregular contact

Fig. 7-9. Discordant dike in high-grade, strongly foliated metamorphic rock with irregular borders. Near Elk City, Idaho.

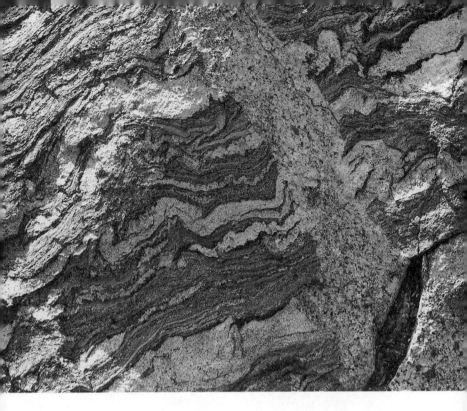

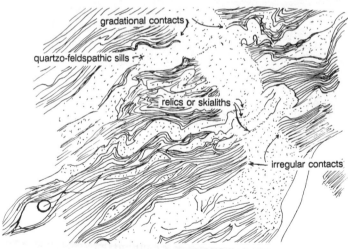

gradational contacts

quartzo-feldspathic sills

relics or skialiths

irregular contacts

Fig. 7-10. Discordant granitic dike in high-grade rocks. Borders are diffused and gradational; some layers of foliation appear to continue across the dike. The dike possibly originated through feldspathization in the solid state rather than intrusion. Near Elk City, Idaho.

Fig. 7-11. Augen gneiss in conformable contact with biotite quartzite. Parallel foliation, defined by biotite, is well developed in both the older quartzite and the younger augen gneiss. Note the thin layers of augen gneiss within the biotite quartzite unit. The conformable contact indicates that the augen gneiss was synkinematic with the deformational event that caused the strong parallel alignment of mica. Bedding in the biotite quartzite also parallels the foliation. The aplite dike and the pegmatite dike at the bottom of the photograph are discordant to the foliation indicating that they are postkinematic. Shoup, Idaho.

248

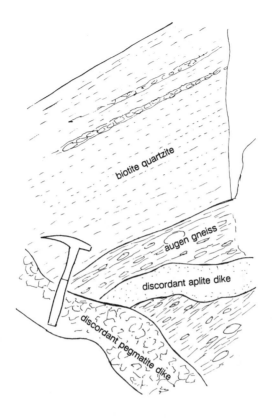

temperatures sufficient for partial melting; (2) imperceptible gradation from pluton to country rock; (3) country rock may be partially melted; (4) migmatite zones; (5) most contacts between country rock and the pluton are concordant; (6) high-grade metamorphic rock; (7) absence of contact metamorphism or chilled contacts; (8) ductile deformation features; and (9) small quartzo-feldspathic ptygmatic veins.

Granitic Magma Generated in the Infrastructure

Anatexis is the melting or breakdown of crustal rocks to form granitic material and migmatite. **Palingenesis** is the derivation of granitic magma from the granitic crust in situ following anatexis. The magma floats upwards like a diapir because the molten granite has a lower density than the surrounding country rock.

Transition Level

The transition level is the highest level of regional meta-morphism and is the most commonly observed emplacement level. Plutons at this level are referred to as **mesozonal plutons** and have the following characteristics: (1) consolidated at 7 to 15 km and at 300 to 600 degrees; (2) primary foliation at margins; (3) combination of regional and contact metamorphism; (4) ductile deformation features; (5) intrusion along the foliation plane such as lit-par-lit gneisses (Fig. 7-11); and (6) general absence of chilled margins because country rock is hot.

Supercrustal (Epizonal) Emplacement

Supercrustal or epizonal emplacement occurs in the brittle upper 5 to 7 km of the crust. It is characterized by the following features: (1) plutons tend to intrude their own volcanics; (2) associated with flexural folds; (3) miarolitic cavities; (4) hydrothermal alteration (Fig. 7-12); (5) breccia; (6) quartz veins; (7) sharp discordant contacts; (8) deformation by brittle fracture; (9) absence of ductile flow (Fig. 7-13); and (10) very little foliation.

Determining Age of Intrusion
Relative to Deformation Event

A common field problem is to determine if an intrusion predates a deformation (prekinematic intrusion), is the same age as the deformation (synkinematic), or postdates the deformation (postkinematic).

A **prekinematic** intrusion is characterized by the following features: (1) intrusion is affected by all later deformational events; (2) schistosity cuts across country rock and intrusions; (3) attitude of intrusion is not related to trends of the deformation event; and (4) grain deformation indicated by strain shadows and granoblastic textures.

A **synkinematic** intrusion is characterized by the following features: (1) intrusion has weaker development of schistosity than country rock; (2) intrusion has weaker development of folding and shearing offsets but same axial trends as country rock; (3) intrusion penetrates along foliation resulting in axial-plane dikes (Fig. 7-14) and **lit-par-lit** injection gneiss (Fig. 7-15) with thin layers of granitic rock alternating with layers of country rock; (4) mineral lineations in both presynkinematic and synkinematic intrusions are parallel to those in the country rock; (5) rotated xenoliths consist of metamorphosed

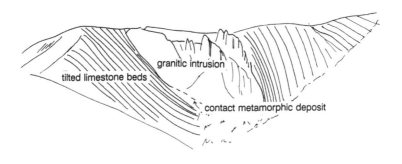

Fig. 7-12. Epizonal granitic pluton pushed up through limestone. The force of the intrusion apparently tilted the limestone beds causing them to be somewhat conformable to the pluton. A well defined metamorphic deposit is formed along the contact with a variety of skarn minerals exposed.

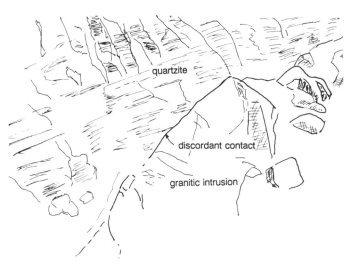

Fig. 7-13. Intrusive contact between an epizonal granitic pluton and quartzite country rock. Note that the contact is discordant without any detectable zonation. South of Oakley, Idaho.

252

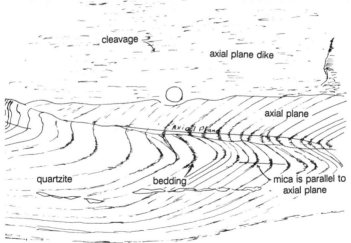

Fig. 7-14. Axial-plane dike intruded along the cleavage plane. Note that the axial-plane cleavage of the folds parallels the dike. Cleavage, formed of parallel biotite grains in the granitic rock, is parallel to the cleavage in the quartzite. This is evidence that deformation of the quartzite occurred sykinematically with the granitic intrusion. Middle Mountain, Idaho.

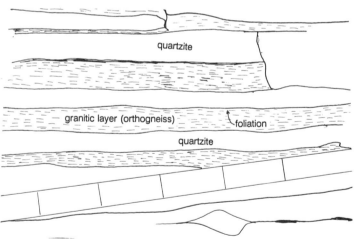

Fig. 7-15. Lit-par-lit gneiss where thin parallel sheets of granitic material penetrated along the cleavage planes of the quartzite. Note that the quartzite layers are more resistant to weathering than the granitic layers. The biotite grains in the granitic rock are parallel to the cleavage of the quartzite. Middle Mountain, Idaho.

country rock (Fig. 7-16); (6) synkinematic intrusions tend to parallel fold axes; and (7) schistosity in dikes and sills parallels cleavage in the country rock.

A **postkinematic** intrusion is is free of deformational features; and associated dikes and contacts tend to be discordant to the metamorphic foliation.

Inclusions

An inclusion in a pluton is any rock type distinct from the general pluton, regardless of origin. **Schlieren** are discontinuous streaks or segregations of dark minerals; they may be partially assimilated minerals or flow banding (Fig. 7-17). **Xenoliths** are inclusions of preexisting rocks. **Cognate xenoliths** (Fig. 7-18) are rocks formed from an earlier stage of the magma. **Skialiths** are remnant inclusions after the surrounding rock has been converted to granitic rock by metasomatic processes. Inclusions may have reaction rings.

Porphyritic Texture

A magma may start to cool under conditions that allow large minerals such as orthoclase or sanidine to form in the early stages of crystallization and later move to a location where cooling is much more rapid (Fig. 7-20). This rapid cooling will freeze the large grains in a groundmass of fine-grained material. The large minerals are called **phenocrysts**. A rock with more than 25 percent phenocrysts is called a **porphyry**. There is much evidence that many so-called phenocrysts grew significantly after complete consolidation of the pluton. In many cases these megacrysts have well developed euhedral form.

Primary-Flow Foliation

Flow layering is generally obvious in obsidian and rhyolite; it is defined by layers of glass shards, crystals and lithophysae. Flow movement in a magma rotates elongate crystals or inclusions parallel to the direction of the flow. Typical minerals include elongate prisms of amphiboles, pyroxenes and books or plates of biotite, plagioclase and orthoclase feldspar.

Platy flow structures are formed by the following features: (1) color contrasts between biotite and feldspar; (2) platy parallelism of crystals; (3) flow layers (schlieren) with tabular aligned aggregates of dark or light minerals from surrounding rock; and (4) xenoliths and segregations parallel to the mineral foliation.

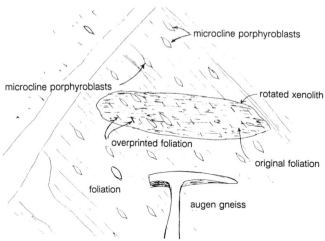

Fig. 7-16. Rotated xenolith of quartzite in augen gneiss. The foliation in the xenolith is parallel to the long direction of the xenolith and is also overprinted by foliation parallel to that of the augen gneiss. This is evidence that the augen gneiss has an igneous origin. Also, the feldspar porphyroblasts in the xenolith are aligned in the same direction as those in the augen gneiss. This is evidence that at least a portion of the augen or porphyroblasts in the augen gneiss were grown in the solid state by metasomatism.

schlieren - layer of dark minerals

Fig. 7-17. Schlieren or segregation of dark minerals in granite. Idaho Batholith.

Fig. 7-18. This ellipsoidal body may be an autolith or a cognate inclusion which is genetically related to the surrounding granitic rock rather than a xenolith derived from wall rock. Wallowa Mountains, Oregon.

Both linear and platy flow structures are common in intrusive and extrusive rocks. The primary foliation forms before consolidation of the magma is complete. **Flow lines** or linear structures generally lie on the foliation plane. **Rift planes** are caused by the tendency of rock to split on flat surfaces formed by microscopic cracks or aligned bubbles and coincide with platy or linear parallelism in rocks.

Volatiles

Rock types which are characteristically rich in volatiles include pegmatite, aplite, orbicular rock and intruded breccia. **Miarolitic cavities** are vesicles caused by volatiles separated

Fig. 7-19. Rounded mafic "pillow" in granitic rock may be derived from mixing of mafic lava with granitic lava. Of course, a petrologic examination would be necessary to confirm the origin. Wallowa Mountains, Oregon.

from the magma. These cavities can be confused with dissolution cavities which are characterized by cutting across grains. Crystals growing in miarolitic cavities are typically euhedral and may include quartz, orthoclase feldspar, epidote, chlorite and carbonates.

Alteration of Plutons

Plutons are altered by hydrothermal waters and magmatic volatiles. Alteration is facilitated by fracturing of the pluton. Because these fractures provide access for hot water to the pluton, you generally find hydrothermal alteration concentrated along fractures in the upper part of the pluton.

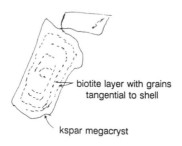

biotite layer with grains
tangential to shell

kspar megacryst

Fig. 7-20. Euhedral phenocryst of microcline in granitic rock. Note the well-defined zoning within the crystal. Near Stanley, Idaho.

SHEET INTRUSIONS

Sheet intrusions include dikes and sills that are generally emplaced after the pluton cools. A **Dike** is a discordant sheet intrusion that cuts across bedding or foliation of the host rock (Fig. 7-21). A **sill** is a concordant sheet intrusion parallel to the layering or foliation of the host rock (Fig. 7-22). Sheet intrusions may create faults to intrude or they may fill preexisting fractures. They tend to be aligned perpendicular to the least compressive stress direction or parallel to the tensile fractures. A set of dikes arranged in an en echelon pattern at the surface may unite and form one continuous dike at depth. A set of parallel dikes, similar to a set of joints, indicates a stress field at the time of intrusion.

Pegmatite Dikes

Pegmatite dikes are rock bodies distinguished from surrounding rock by relatively coarser grain size. Igneous pegmatites, found in large granitic plutons, form from the residual (last to crystallize) volatile-rich fractions of the magma. Pegmatites crystallize at a relatively low temperature, generally between 250 and 700 degrees C.

Zoning. Pegmatites tend to be zoned (Fig. 7-23), and in some cases they may have up to four zones. The outer or border zone is typically less than 10 cm thick and has an aplitic texture. The texture or grain size coarsens towards the core. In the intermediate zone(s) minerals such as plagioclase, perthite, quartz, muscovite, tourmaline, biotite, apatite, beryl, garnet, uranium, thorium, lithium, niobium, tantalium and rare earths may be found. The core is characterized by large amounts of massive white quartz, orthoclase, tourmaline, topaz, beryl and is the best place to search for valuable minerals. A gas cavity may occur at one or more places along the centerline of the pegmatite. Minerals in a gas caity that project inward may have well-developed crystal faces and be valuable as mineral specimens (Fig. 7-24).

Occurrence. Pegmatites tend to be larger and more numerous at the roof of plutons than at any other part of the pluton. Because their distribution and attitudes are controlled by fracture systems, they may intrude the country rock surrounding the pluton. Pegmatites carrying the most valuable minerals tend to occur at the roofs of plutons. Simple pegmatites, without valuable minerals, generally occur in the deep interior of plutons.

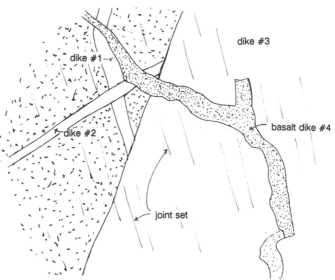

Fig. 7-21. Intersecting dikes of various ages. Relative ages can be established by cross-cutting relationships. The small basalt dike is youngest because it cuts across all the other dikes. Near Lowman, Idaho.

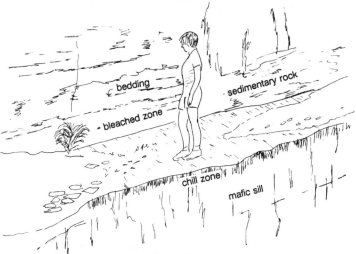

Fig. 7-22. Diorite sill injected along bedding plane of Helena Formation of the Precambrian Belt Supergroup. The upper meter of the sill has a fine-grained chill zone indicating more rapid cooling and crystallization than the central part of the sill. The sedimentary rock near the contact is bleached and altered from the heat.

263

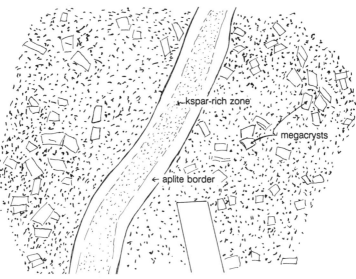

Fig. 7-23. Pegmatite dike has well defined zoning. Note that this rock is composed of about 30 to 40 percent large euhedral phenocrysts. Near Lowman, Idaho.

264

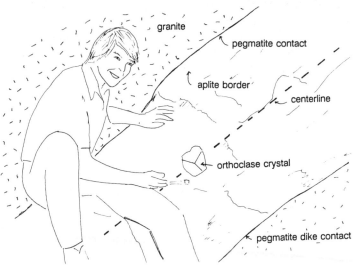

Fig. 7-24. Pegmatite dike in granite about 1 m thick. Note the large euhedral crystal of orthoclase along the centerline of the pegmatite. Silent City of Rocks, south-central Idaho.

265

Fig. 7-25. Pegmatite or graphic granite with intergrowths of quartz and orthoclase feldspar. North of Boise, Idaho.

Evidence of Volatiles. Pegmatites rich in volatiles are characterized by the following evidence: (1) large grain size; (2) euhedral feldspar crystals; (3) segregation of minerals; (4) alteration of earlier-developed minerals; (5) vug-filling by late-stage minerals; and (6) mineral dissolution.

Graphic texture is caused by intimately intergrown grains of potassium feldspar and quartz. The overall pattern suggests the wedge-shaped characters in the writing of ancient Assyria, Babylonia and Persia (Fig. 7-25).

Aplite Dikes

Aplite dikes are common in intermediate to silicic rocks. They are light colored, fine grained and consist primarily of sodic plagioclase and quartz in the groundmass and, in some cases, orthoclase phenocrysts. Like pegmatites, aplite dikes may represent a residual fraction of silica-rich magma after most of the magma has crystallized. This may explain why both pegmatite and aplite dikes tend to occur together.

Basalt Dikes

Basalt dikes are common in plutonic rocks and, as a general rule, postdate the pegmatite and aplite dikes (Figs. 7-26 and 7-27). They tend to follow the fractures already

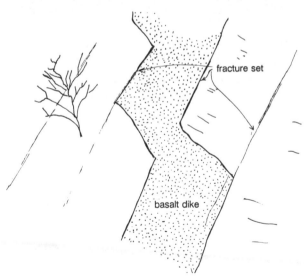

Fig. 7-26. Basalt dike switched fractures at mid photo. Near Lowman, Idaho.

Fig. 7-27. Basaltic dike intruded into tuffaceous volcanic rock. Columnar jointing is well developed with columns perpendicular to the cooling surface. Notice how the basaltic dike rock is much more resistant to weathering and erosion at the surface than the surrounding tuffaceous rock. Near Challis, Idaho.

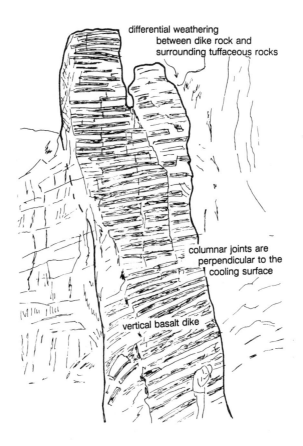

differential weathering
between dike rock and
surrounding tuffaceous rocks

columnar joints are
perpendicular to the
cooling surface

vertical basalt dike

developed in the pluton (Fig. 7-26). If basalt dikes reach shallow levels in the crust where the rocks have lower temperatures, they may have columnar joints developed perpendicular to the cooling surface (Fig. 7-27).

Orbicular Rocks

Orbicular rocks are ellipsoidal-shaped masses of rock consisting of successive shells of dark minerals (biotite) and light minerals (feldspar). The occurrence of orbicular rocks is a rare phenomenon. There are fewer than 200 known localities throughout the earth. The State of Idaho happens to have at least three of these localities: (1) one in the Buffalo Hump area in central Idaho (Figs. 7-28 and 7-29), (2) one in southwest Idaho near Banks (Fig. 7-30), and (3) one near Shoup in east-central Idaho. The orbicular rocks near Shoup crop out for about 2 km along the south side of the Salmon River.

Fig. 7-28. Densely clustered orbicules in Buffalo Hump Mining District, Idaho. Note the nature of the border contacts among these orbicules. They appear to have deformed in a ductile manner to accommodate their neighbors.

Orbicular Rocks Near Shoup, Idaho

The shape of the intrusion containing the orbicular rocks is very irregular. Along its periphery, numerous dikes of quartz diorite interfinger and discordantly penetrate the augen gneiss country rock (Fig. 7-31). Evidence in the field is persuasive for a dynamic emplacement of the intrusion. More than 50 percent of the total volume of the intrusion consists of angular xenoliths, xenocrysts and autoliths in a medium-grained, quartz diorite matrix. In other parts of the intrusion, the quartz diorite matrix represents up to 90 percent of the rock volume. Primary-flow foliation and schlieren tend to give the intrusion a gneissic appearance.

Breccia fragments are primarily xenoliths of augen gneiss, quartzite, biotite gneiss and biotite schist. Some of the xenoliths may have been transported a long distance because they are dissimilar to the enclosing rock types. The size range of the xenoliths is variable, with some blocks of augen gneiss almost 100 m in diameter.

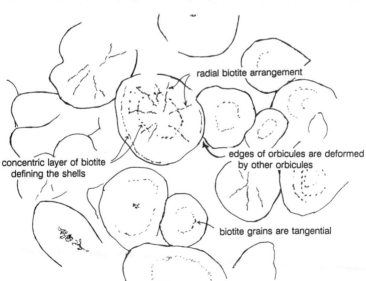

Fig. 7-29. Orbicular granitic rock near the Buffalo Hump Mining District, central Idaho. The orbicules have both radial and concentric arrangement of biotite grains.

271

Fig. 7-30. Orbicular granitic rock exposed in a road cut near Banks, Idaho.

The orbicules, which occur in clusters, were formed by the crystallization of alternate layers of plagioclase and biotite around the nucleus. However, in some cases nucleation occurs around xenocrysts and autoliths. Typically, the orbicules have a nucleus of coarse-grained biotite schist. These biotite schist xenoliths probably were brought up from deep in the crust because they are different from any rocks in the area. Although most xenoliths are mantled by at least one layer of plagioclase, many of the large angular xenoliths have several shells of biotite and plagioclase. The single plagioclase mantles are found on xenoliths of all rock types. Orbicules have up to 10 shells of plagioclase with each shell 3 mm to 1 cm thick (Fig. 7-32). The individual orbicules generally have a sharp contact with the surrounding matrix. For the most part, the external shape of the orbicules depends on the shape of the xenolithic nucleus. Shapes vary from spherical to ellipsoidal masses. Fragments of orbicule shells indicate that some orbicules may have been brittle at the time of emplacement; however, other orbicules were apparently deformed in a ductile condition as they were blasted against the host rock.

272

Fig. 7-31. Granitic intrusion, with well developed orbicules occurring in clusters, cutting high-grade gneissic rocks near Shoup, Idaho.

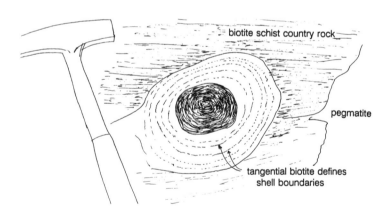

Fig. 7-32. Large, well developed orbicule about 20 cm in diameter with many alternating shells of biotite and plagioclase. The nucleus is a large black mass of coarse biotite with grains in concentric arrangement. The biotite grains in the shells are tangential to the surface of the spheroidal shells. Near Shoup, Idaho.

Fig. 7-33. Spheroidal weathering in granitic rock. Layer by layer exfoliation through the process of granular disintegration causes most granitic rock exposures to evolve towards rounded forms. Silent City of Rocks, Idaho.

Field Identification of Granitic Rocks

Granitic outcrops are easily recognizable in the field, even from distances of several hundred meters. Under close inspection, granite has a salt and pepper appearance with dark minerals of biotite mica and hornblende and light minerals of plagioclase and quartz. The constituent minerals can generally be identified without a hand lens. Of course, many minor accessory minerals are too small to be identified with the unaided eye.

The coloration of outcrops tends to be very light gray to very light tan. Outcrops are generally smooth and rounded (Fig. 7-33) as a result of surface weathering by granular disintegration and spalling (Fig. 7-34). Most exposures are cut by one or more sets of fractures which may give the outcrop a blocky appearance.

275

Fig. 7-34. Spheroidal weathering of granitic rock by exfoliation or spalling of thin layers less than 1 cm thick. Silent City of Rocks, Idaho. Spheroidal weathering leads to rounded forms which are very common in granitic rocks and somewhat less common in volcanic and sedimentary rock. Chemical weathering transforms feldspars to clay. Because clay absorbs water, the volume of the original feldspar expands by a process called granular disintegration. In order for chemical weathering to occur, water must have access to the feldspars so feldspar near the surface or along joints is altered to clay. As the near-surface altered feldspar expands, the expansion disintegrates or breaks up the interlocking mineral grains towards the exposed face and thin flakes spall off. Movement of water along fracture planes also causes feldspar to progressively decompose from the outside towards the interior. So, each successive shell towards the interior becomes more and more rounded.

The Silent City of Rocks

The "Silent City of Rocks," a 25-square-km area in Cassia County, Idaho, embraces a remarkable variety of rock forms created by a combination of jointing, weathering and case hardening. These rock forms are composed of a 28 million-year-old granite. Significantly, a shallow emplacement of this granitic pluton is indicated by a lack of foliation at the margins and discordant contacts with the country rock.

Jointing. Three well developed joint sets establish the basic forms in the City of Rocks. Joints facilitate the weathering processes by providing a plumbing system for solutions to migrate into the outcrops to cause the alteration and disintegration of surface layers of granite (Fig. 7-35). These large fracture channels for fluids make it possible for blocks to separate and form tall, isolated monoliths such as spires and turrets.

Weathering. Although jointing controls the general form of outcrops in the City of Rocks, weathering is the agent responsible for creating the bizarre and fantastic shapes that characterize the area. On the surface of outcrops, weathering occurs by granular disintegration. Chemical weathering occurs by solutions which penetrate the cleavage cracks in crystals and between mineral grains. Once the solutions are in these narrow boundaries, they cause clay to form from feldspar which has a larger volume than the space available. This process of hydration and other chemical changes cause the disintegration and spalling. In other words, one layer of crystals after another is successively removed from the surface. This leaves the newly exposed surface in a smooth, rounded condition with no sharp edges or corners. The detrital material **grus** weathers from the granite and is carried by wind and water to low areas among the prominent forms (Fig. 7-36). The grains of quartz, feldspar and mica at the surface of outcrops are friable and easily disintegrated with hand tools.

Case Hardening. In addition to granular disintegration, case hardening is important in developing the unusual erosional forms. At most of the outcrops, an outer layer has been hardened by the deposition of minerals such as iron oxides. Once a form has acquired a case-hardened protective shell, the soft inner granite is protected until the shell is breached. When the interior granite is accessible to wind and water, caves, niches, arches, bath tubs or sinks, toadstools and

277

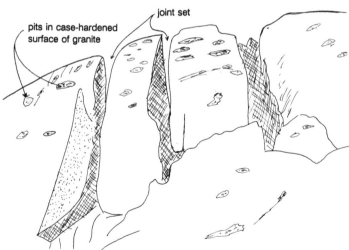

pits in case-hardened
surface of granite

joint set

Fig. 7-35. Weathering has developed wide spacing at the joints. The joints facilitate the weathering process by allowing access of fluids along the joints so that granular disintegration can break up the material in the joint spaces. Silent City of Rocks, Idaho.

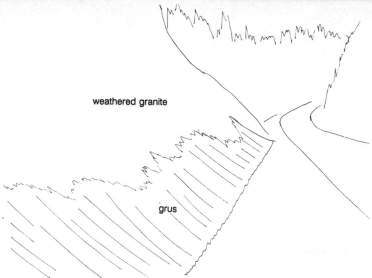

weathered granite

grus

Fig. 7-36. Roadcut through granitic rock. Note the apron of grus at the toe of the slope. This coarse sand forms from weathering of granite by granular disintegration. Near Lowman, Idaho.

hollow boulders are formed by a combination of chemical weathering and erosion (Figs. 7-37 and 7-38). In some cases only the case-hardened shell is left.

Fig. 7-37. Surface of case hardened granite has numerous **weathering pits** formed when weathering processes break through the relatively thin hardened surface and excavate the pits by granular disintegration and spalling. During periods of precipitation, these pits fill with water which accelerates weathering. The loose granular material, **grus** released in the pit may be blown out during windy dry periods.

Cavernous weathering is responsible for the caves, arches, hollow forms, niches and alcoves characteristic of granitic rocks. This type of weathering tends to occur in rocks that are (1) vulnerable to case hardening by iron and manganese oxides, and (2) granular disintegration.

Fig. 7-38. Granitic outcrop with case hardened rounded surface. Once the shell is penetrated, the soft inner core can be removed by the action of water and wind. Silent City of Rocks, Idaho.

MINERAL DEPOSITS ASSOCIATED WITH INTRUSIVE ROCKS

Hydrothermal Deposit

A hydrothermal deposit is one precipitated from a high temperature solution. As hot water with minerals in solution rises towards the earth's surface, the lower temperature and pressure near the surface cause the minerals to precipitate out of solution.

Preparation of Rocks for Mineralization

Several processes must affect the rock in order to make it more receptive to mineralization. The rock must become more permeable and brittle. Rocks are hardened by silica, then shattered by faulting so as to increase permeability. Broken silica has clean fractures with little or no powder so that fluids can move easily through the rock. Typically, rocks with a high porosity such as sandstones and conglomerates also have high permeability. Shales, on the other hand, have a high porosity but a low permeability. Consequently, shale beds may confine and trap a mineralizing fluid rather than allow it to pass through. Joints and contraction cracks in igneous rock make excellent channelways for fluids. Vesicular layers and inter-beds between lava flows also provide very good permeability.

Fractures and Mineralization

A **shear zone** is a highly fractured zone with closely spaced, subparallel fault planes. Theoretically, shear fractures should not be opened because movement is parallel to the plane of the fracture; however, shear surfaces are normally curved so that movement produces alternating open and closed spaces. Although tension fractures are open because of tension normal to the walls, they are also opened by movement oblique to the plane of the fracture. So a shearing component may help to open them.

Faults formed near the surface are generally more open and have higher permeability than deep faults. Thrust faults are caused by compression and generally have a fault plane that dips 30 degrees or less to horizontal. Thrust faults have tight fractures containing much gouge, low permeability and are poor for mineralization. Gravity or normal faults are caused by extension; they tend to be open, permeable and excellent for mineralization. The fault planes of normal faults tend to dip 40 to 70 degrees.

Fault Gouge

Gouge zones are produced where faults move along uneven surfaces. Gouge, which is clay-sized, finely ground rock, greatly restricts the access of mineralizing solutions. Brittle quartzites make either clean breaks or shattered zones and have high permeability; shales and many igneous rocks make tight fractures with much gouge and have a low permeability.

Veins tend to narrow where they cut granitic rock, because it powders during faulting. Feldspar in the granite alters to clay upon exposure to hydrothermal solutions. Therefore, a fault in granitic rock can cause reduced permeability in two ways: (1) by grinding of rock to clay-sized particles; and (2) by altering feldspar to clay minerals which take up greater volume.

Hydrothermal Veins

Hydrothermal veins form by filling of open fractures or replacement, or both, along open fractures (Fig. 7-39). Vein filling may include a variety of minerals such as sulphides, quartz, carbonates, fluorite (Fig. 7-40) and oxides. Ore is commonly localized at fracture intersections where permeability is particularly high.

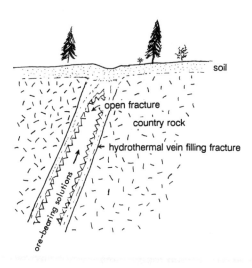

Fig. 7-39. Cross section of an open fracture partially mineralized by ore-bearing solutions. If ore-bearing solutions reach sufficiently high levels in the crust where the temperature and pressure are low enough to allow the minerals to precipitate, a mineral deposit is formed.

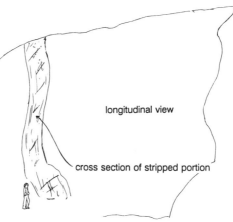

Fig. 7-40. Longitudinal view of fluorite vein exposed west of Challis, Idaho. Note the three people viewing the vein at the lower left of photograph. The hanging wall of the vein apparently broke away longitudinally along the centerline. Fluorite crystals and botryoidal structures cover a large portion of the exposed surface of the vein, indicating crystal growth into an open cavity.

284

Pipes or Chimneys are rod-shaped bodies with a definite plunge which may form at the intersections of two vertical faults. They normally lie at a steep angle.

Breccia pipes typically form at the intersection of two tabular features such as fractures, dikes and bedding.

Diatremes are volcanic explosion vents formed where gases under high pressure explode upwards. Many diatremes cause breccia pipes. They are thought to form by explosive fragmentation of rock into breccia by superheated steam. Diatremes form in areas of igneous activity which can subsequently supply hydrothermal mineralizing fluids to these exceptionally porous and permeable rocks. Evidence for a pressurized, pneumatolytic origin include vesicles and well-rounded grains.

Saddle Reefs form in the hinge areas of folds. Where competent layers of rock such as sandstone are interlayered with incompetent rock such as shale, compression will occur at the limbs and tension at the hinge areas. Consequently, tension fractures parallel to the axial plane may develop in the crestal area of competent rock. Mineralizing fluids can then move in along the permeable and porous crests of the folds.

Stockworks are vertical pipes caused by shattering of igneous rock into a network of interconnecting fine cracks that may be mineralized.

Mineralization at Contacts

Contacts between igneous intrusions and country rock are commonly sites for ore deposition because a massive intrusion will fracture the surrounding rocks particularly at the epizonal level. Furthermore, the igneous intrusion provides the essential mineralizing solutions.

Chemical Controls

Factors that cause minerals to precipitate out of hydrothermal solution include reductions in temperature and pressure and reactive wall rock. Solubility tends to increase in proportion to increasing temperature. Minerals are precipitated if a solution is saturated for a given temperature. In other words, as the fluid migrates up towards the surface, the temperature drops and minerals are precipitated. Similarly, mineral-bearing fluids, when moving upward and into open spaces may deposit minerals because of the loss of pressure. Also, as these fluids move into new areas the presence of ground water or more reactive rocks may cause minerals to precipitate. Carbonates, for example, are permeable and chemically reactive with weak acids; consequently, they are selectively replaced by mineral solutions.

Fig. 7-41. Hydrothermal, subvertical quartz vein in quartzite country rock.

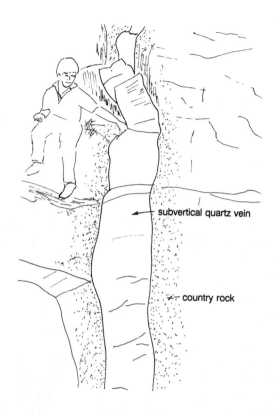

subvertical quartz vein

country rock

Epithermal Deposits

Epithermal deposits are an important source of lode gold deposits. They are formed at less than 1,000 m from the surface and at low temperatures ranging between 50 and 200 degrees centigrade. Mineralization occurs by open-space filling with such textures as drusy (crystal lined) cavities, symmetrical banding and comb structures. The fissures may open at the surface as hot springs. Epithermal veins are typically related to Tertiary plutons and volcanism.

Predicting Gold Value with Depth

Gold is an inert, insoluble mineral and is not readily susceptible to leaching. As a result, the gold content of the rock may decrease with depth. Another depth problem occurs if gold is contained in sulfides. In such a case, free gold is only available in the oxide zone or the zone of weathering above the water table.

Ore Fluids

As hot fluids are discharged from magma, they may circulate through huge volumes of fractured rock dissolving a variety of minerals. After taking minerals in solution at high temperatures and pressures, the fluids move towards the surface along permeable channels such as fracture zones. When temperature and pressure drop sufficiently, minerals will begin to precipitate along the walls of the fractures.

Jasperoid

Jasperoid is a cryptocrystalline silica that is transported in hydrothermal solutions and replaces country rock. It is normally deposited as resistant masses along faults. If jasperoid is fractured, it may be almost free of clay and provide increased permeability. Jasperoids are relatively inert to ore-forming fluids and free of gouge and altered wall rock.

Quartz Veins

Only a small percentage of vein quartz will contain gold. **Bull quartz** is a term for a glassy quartz that is generally barren of gold (Fig. 7-41). Gold below the oxide zone is generally associated with sulfides. Ore minerals with gold in sulfide include pyrite, chalcopyrite, arsenopyrite and galena; however, gold may also exist in a free state below the oxide zone. Iron streaks and vugs lined with rusty crystals in quartz veins are promising for gold (Fig. 7-42). If gold is present in such veins, it may be possible to see it with a hand lens or the naked eye.

In a shear zone the layers of rock may be replaced by sulfide and/or quartz. This may produce a "laminated quartz" or "ribbon texture." Clear white quartz may represent open cavity filling; and gray or green quartz with inclusions may represent replacement material.

Hydrothermal Alteration

Hydrothermal alteration is caused by hot water moving through fractures and permeable zones. A gas or steam phase with dissolved material may also enhance the alteration. These fluids are either exsolved magmatic water or meteoric groundwater heated by the pluton and circulated by convection. Carbonate and glassy igneous rocks are particularly susceptible to alteration. Where one significant mineral is added, these rocks are referred to by such terms as silicified (Fig. 7-43), dolomitized, cloritized and albitized. Veins may be deposited by open-space filling, replacement, or both. Typically, the wall rocks enclosing these veins are hydrothermally altered.

Fig. 7-42. Rusty quartz vein almost 1 m thick exposed by shaft. The vein has been fractured and mineralized with iron and possibly gold. Buffalo Hump Mining District, Idaho.

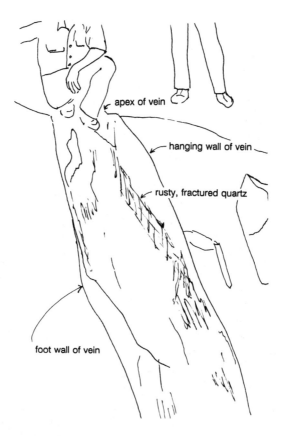

apex of vein

hanging wall of vein

rusty, fractured quartz

foot wall of vein

Open-Space Filling

Open-space filling occurs at shallow levels in the earth's crust where rocks fracture rather than flow in response to deformation. Open-space filling can take place much more rapidly than replacement processes because of the free access of mineralizing solutions. Evidence for open-vein filling includes the following: (1) vugs and cavities (Fig. 7-44); (2) minerals become coarser towards the center (Fig. 7-45); (3) vein may be zoned by successive layers of minerals of different composition in response to changing composition of mineralizing fluids; (4) comb structure or terminated crystals grow towards the center line (Figs. 7-46, 7-47 and 7-48); (5) composition and texture on one side may be symmetrical to that of the other; (6) oblique structures cut across the veins indicating that the vein dilated; and (7) opposite walls of the vein should match indicating they were originally joined.

290

Fig. 7-43. Portion of exceptionally large quartz vein in the Buffalo Hump Mining district of central Idaho. The vein, which is almost 100 m thick and 1500 m long, consists of anastomosing, hydrothermal quartz veinlets in granitic rock.

Replacement

Evidence that replacement has occurred can be indicated by the following criteria: (1) crystals cut across original minerals and structures; (2) vugs may be caused by solution from replacement processes rather than open-space filling; (3) one mineral may be selectively replaced over another; (4) gradational contacts may indicate replacement, whereas, abrupt contacts may be caused by open-vein filling; (5) inclusions of wall rock are surrounded by replaced material; (6) pseudomorphs or crystals of replaced material fill the crystal form of the original mineral; and (7) the two walls of a fracture do not match.

Fig. 7-44. Shattered dark country rock has thin veinlets of fluorite and cavities lined with fluorite. In these cavities you can find fluorite cubes where crystal growth was unimpaired.

Gangue

Gangue includes those minerals that have no value in an ore deposit. They are commonly the silicate and carbonate minerals. In cases where the gangue minerals have the same genesis as the ore minerals, gangue minerals could be used as a guide to new ore. For example, smoky quartz may indicate uranium, fluorescent calcite may indicate ore nearby and dark purple fluorite may be associated with uranium minerals.

Gossans and Leached Outcrops

A gossan or iron hat is a porous, rusty capping on a sulfide deposit. Any outcrop or float of iron-stained, light-colored igneous rock, fractured and recemented with silica, should be carefully examined because gossans and leached cappings have long been an indication of ore. Sulfides are unstable under surface conditions. They leave several hydrous iron oxides, generally referred to as "limonite," and in some cases gold as

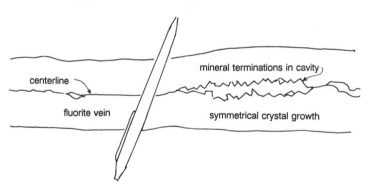

centerline

fluorite vein

mineral terminations in cavity

symmetrical crystal growth

Fig. 7-45. Small fluorite vein shows symmetrical crystal growth from the vein towards the centerline. Near Challis, Idaho.

293

Fig. 7-46. Fluorite cubes formed by crystal growth towards an open fracture. Near Challis, Idaho.

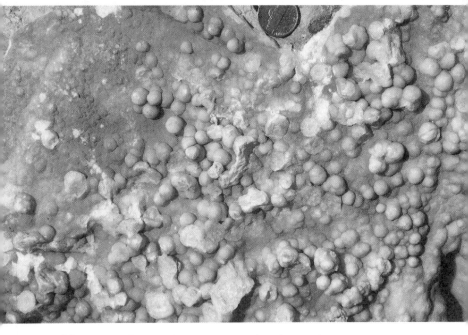

Fig. 7-47. Botryoidal structures (surface of spherical shapes) provide evidence of open-fracture filling at fluorite mine. Near Challis, Idaho.

the only valuable mineral. In certain cases, the color and texture of the limonite can indicate the original sulfide minerals. Although most exposed gossans have been examined, there are undoubtedly many more which are obscured by overburden and (or) vegetation.

Deposits Formed by Secondary Enrichment

Some mineralized vein deposits are enriched at or below the water table by a process called supergene enrichment. Surface water moving along the fractures above the water table in the zone of oxidation dissolves minerals and carries them in solution down to the water table (Fig. 7-48). At the water table, secondary minerals are deposited which are generally much richer than primary minerals originally deposited in veins. For example, a typical primary sulfide mineral is chalcopyrite with 34.5 percent copper. If this mineral is taken into solution and carried down to the water table, the copper may again be deposited in the form of bornite (63 percent copper), covellite (66 percent copper) or chalcocite (80 percent copper). The following minerals are commonly found in gossans or oxidized upper portions of veins:

Iron minerals - rusty brown, yellow, red
Copper minerals - blue, green
Nickel ores - pale green
Cobalt - pink, red color
Molybdenum - pale yellow
Manganese - sooty black
Uranium - bright orange, yellow, green

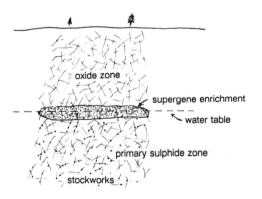

Fig. 7-48. Cross section of a mineral deposit formed by supergene enrichment.

Skarn and Contact Metamorphic Deposits

After intrusion, a magma gives off heat and fluids. These hot fluids migrate towards low temperature and pressure. New minerals and textures form along the contact of the pluton and the country rock (Figs. 7-49 and 7-50). If the country rock is a limestone, it may be recrystallized into a marble along the contact. The intruded magma supplies valuable metals and silica. Silica precipitates in the pores of sedimentary rock as a quartz cement and reacts with chemicals in the country rock to form silicate minerals. Hot solutions leach out portions of the country rock and in its place silica and other minerals are deposited. Deposition occurs in permeable beds along bedding planes, cavities and fractures. Metals are very mobile and tend to be driven out of the magma and localized in the roof of the magma chamber.

Skarn minerals are formed at the contact between a granitic pluton and a carbonate-rich rock such as a limestone. Silicon, Al, Fe and Mg are introduced, possibly from an intrusive source into calcium and magnesium carbonates (limestone and/or dolomites). Skarn minerals include grossularite garnet, scheelite, wollastonite, epidote, biotite, corundum, quartz, diopside, tremolite, fluorite, sphalerite, tourmaline and topaz.

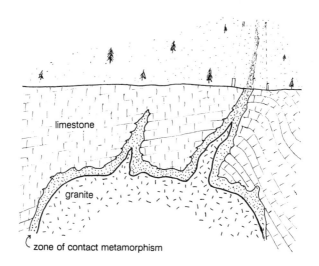

Fig. 7-49. Cross section of granitic intrusion in limestone. Along the "lime-granite contact" as the miners called it, a zone of metamorphism commonly produced skarn minerals such as garnet, quartz, epidote and some ore minerals.

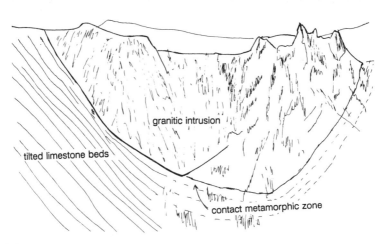

Fig. 7-50. Small epizonal granitic pluton pushed up through the limestone. Note how the intrusion caused the beds of limestone to be tilted to the extent they are almost concordant to the border of the intrusion. Skarn minerals derived from contact metamorphism may be observed along the limestone-granite contact. McKay Mining District, Idaho.

297

Pegmatite

Pegmatites are very coarse-grained igneous or metamorphic rocks. Igneous pegmatite forms from residual volatile-rich fractions of the magma; whereas, metamorphic pegmatite forms by mobile constituents that concentrate during metamorphic differentiation. Pegmatites may have a tubular or dike-like shape or may be lensoid masses. They are generally less than a meter thick, but may exceed 100 m in width and have a length measured in hundreds of meters.

Although some mafic pegmatites are known, most pegmatites have a silicic to intermediate composition. Pegmatite dikes are generally found in and near the roofs of large plutons (Fig. 7-51). Most of these dikes have a very simple mineralogy. Typical minerals include quartz, orthoclase feldspar and mica. Red garnets and black tourmaline are also common as small disseminated crystals. Many valuable economic minerals as well as crystal specimens are recovered from pegmatite. These minerals include quartz, feldspar, micas, chalcopyrite, molybdenite, sphalerite, beryl, apatite, tourmaline, monazite, topaz, garnet, spodumene, cassiterite and lepidolite. Rare earth minerals found in pegmatite include tantalum, niobium, beryllium, lithium, cesium, uranium, cerium and thorium.

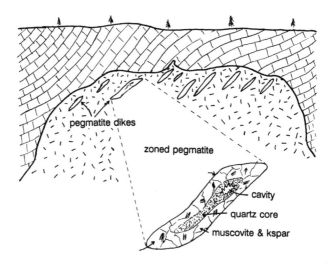

Fig. 7-51. Cross section of a granitic intrusion. Notice how pegmatite dikes tend to form in the upper portion of the pluton.

Most pegmatite dikes are characterized by a crude zoning. This happens because a pegmatite crystallizes somewhat like a geode, from the outside towards the center. Pegmatite dikes typically have a quartz core because quartz is generally one of the last minerals to crystallize.

Some pegmatite dikes have a gas cavity along the centerline of the dike. These cavities range from 2 cm to more than 0.5 m in length and may contain large crystals with fully developed crystal faces. Gem minerals such as amazonite (green microcline) topaz, beryl (aquamarine) and smoky quartz are common in pegmatite.

Prospecting for Pegmatite

One of the best ways to find pegmatite dikes with pockets or cavities in which crystal specimens may be found is to carefully examine the float. **Float** is a term used to describe fragments of a pegmatite deposit that might be detached and moved downslope. Look for large pieces of quartz with attached crystals of amazonite and tourmaline. Also, large pieces of feldspar and mica indicate a pegmatite. Crystals with faces are especially diagnostic because they indicate a pocket exists in a pegmatite where other crystals may be found. Pegmatites form low areas because they tend to weather relatively quickly; as a result, vegetation may thrive over a pegmatite dike. However, the quartz core is more resistant that the surrounding minerals and will stand out in high relief. This quartz may be rose, gray, smoky or amethyst capped. Pegmatites generally do not occur as a single dike but rather as a group of dikes. So, if one is found, there may be more within 50 to 100 m.

Mineral Identification

Books of muscovite mica are very common in pegmatite. These books tend to increase in size towards the centerline of the pegmatite body. Pink, lithium-rich mica is called lepidolite. Orthoclase feldspar is very commonly found as large flesh-colored crystals. Beryl crystals are generally found imbedded in quartz; they tend to range from pale green to blue in color. Tourmaline occurs as long, black, rod-like crystals which generally point towards the center. Translucent white quartz commonly occurs in the core of pegmatite.

8 Metamorphic Rocks

Metamorphic rocks are those that have beem transformed from preexisting rock into texturally or mineralogically distinct new rocks by high temperature, high pressure or chemically active fluids. One or more of these agents may induce the textural or mineralogical changes. For example, minute clay minerals may change into coarse mica. Heat is probably the most important single agent of metamorphism. Metamorphism occurs within a temperature range of 100 to 899 degrees centigrade. Heat weakens bonds and accelerates the rate of chemical reactions. Two common sources of heat include friction from movement and intrusion of plutons. Pressure changes are caused primarily by the weight of overlying rock. Where there are more than 10,000 m of overlying rock, pressures of more than 40,000 psi will cause rocks to flow as a plastic. Pressure may also be caused by plate collision and the forceful intrusion of plutons.

Chemically active fluids (hot water solutions) associated with magma may react with surrounding rocks to cause chemical change. Directed pressure is pressure applied unequally on the surface of a body and may be applied by compression or shearing. Directed pressure changes the texture of a metamorphic rock by forcing the elongate and platy minerals to become parallel to each other. Foliation is the parallel alignment of textural and structural features of a rock. Mica is the most common mineral to be aligned by directed pressure.

Types of Metamorphism

There are two types of metamorphism: contact metamorphism and regional metamorphism. Contact metamorphism is the name given to country rock which reacts with an intrusive rock. Changes to the surrounding rock occur as a result of penetration of the magmatic fluids and heat from the intrusion. Contact metamorphism can greatly alter the texture of the rock by forming new and larger crystals. In contact metamorphism, directed pressure is not involved so the metamorphosed rocks are not foliated.

Most metamorphic rocks are formed by regional metamorphism. Metamorphic rocks are typically formed in the cores of mountain ranges, but may be later exposed at the surface by erosion. Typical rock types include foliated slates, phyllites, schists and gneisses.

Common Metamorphic Rocks

Marble is a coarse-grained rock consisting of interlocking calcite crystals. Limestone is recrystallized during metamorphism into marble.

Quartzite forms by recrystallization of quartz-rich sandstone in response to heat and pressure. As the grains of quartz grow, the boundaries become tight and interlocking. All pore space is squeezed out; and when the rock is broken, it breaks across the grains. Quartzite is the most durable construction mineral. Although both marble and quartzite may be white to light gray, they may be readily distinguished because marble fizzes on contact with dilute hydrochloric acid whereas quartzite does not. Also, marble can be scratched with a knife whereas quartzite can not.

Slate is a low-grade metamorphic equivalent of shale. It is a fine-grained rock that splits easily along flat, parallel planes. Shale, the parent rock, is composed of submicroscopic platy clay minerals. These clay minerals are realigned by metamorphism so as to create a slaty cleavage. In slate, the individual minerals are too small to be visible with the naked eye.

Phyllite is formed by further increase in temperature and pressure on a slate. The mica grains increase slightly in size but are still microscopic. The planes of parting have surfaces lined with fine-grained mica that give the rock a silky sheen.

Schist is characterized by coarse-grained minerals with parallel alignment. These platy minerals, generally micas, are visible to the naked eye. A schist is a high-grade, metamorphic rock and may consist entirely of coarse, platy minerals.

Gneiss is a rock consisting of alternating bands of light and dark minerals. Generally the dark layers are composed of platy or elongate minerals such as biotite mica or amphibolite. The light layers typically consist of quartz and feldspar. A gneiss is formed under temperatures and pressures sufficient to cause minerals to segregate into layers. A gneiss is formed at temperatures slightly below the melting temperature of rock. If temperatures become sufficiently high, the rock begins to melt and magma is squeezed out into layers within the foliation planes of the solid rock. The resulting rock is called a **migmatite** - a mixed igneous and metamorphic rock.

Granitic Stringers

Granitic stringers are common within some metamorphic rocks, particularly high-grade gneisses (Fig. 8-1). If possible, determine if they are concordant or discordant to the foliation

Fig. 8-1. Ptygmatic folds of quartzo-feldspathic layers in schist. These veins are both concordant and discordant to the foliation of the schist.

and related to a deformational event. Granitic stringers are typically 1 to 4 cm thick, but can be much thicker. Boudinage and tight, polyclinal ptygmatic folds are common to the granitic stringers. The enclosing foliation may also be folded with the granitic stringers. Many granitic stringers have a concentration of biotite at the edges.

Granitic stringers tend to have a simple mineralogy consisting of plagioclase, quartz and biotite. The texture is invariably crystalloblastic. Most are formed by one of two different mechanisms, or a combination of these mechanisms: (1) **lit-par-lit** injection (Fig. 8-2), and (2) **metamorphic differentiation** (Fig. 8-3). In lit-par-lit gneiss, the quartzo-feldspathic layer is injected along the foliation plane during the deformational event that generated the foliation (synkinematic intrusion).

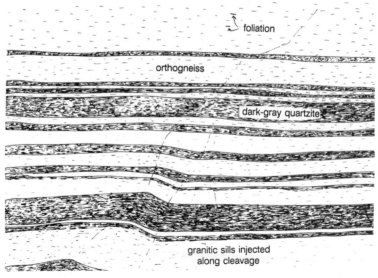

Fig. 8-2. Lit-par-lit gneiss with granitic material injected along the cleavage layers in dark gray quartzite. The granitic layers have secondary foliation of aligned mica grains parallel to the quartzite. Middle Mountain, Idaho.

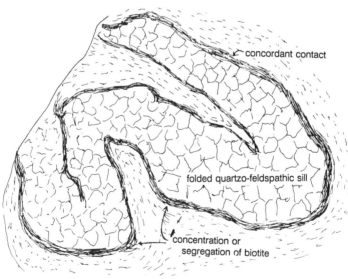

Fig. 8-3. Folded quartzo-feldspathic layer concordant to strong schistosity. Note that at the margin of the granitic layer there is a biotite-rich layer possibly suggesting an origin by metamorphic segregation where the biotite at the margin may have been derived from the quartzo-feldspathic layer. Shoup, Idaho.

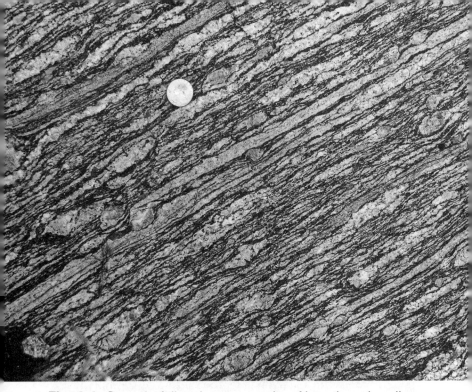

Fig. 8-4. Strongly foliated augen gneiss. Note the microcline porphyroblasts are flattened and elongated along the plane of foliation. Near Elk City, Idaho.

Metamorphic Differentiation. The concentration of large biotite grains at the edges of the quartzo-feldspathic layers is indicative of this mechanism.

Augen Gneiss of Central Idaho

The augen gneiss in central Idaho may be a large 1,500 million year old plutonic complex of batholithic proportions originally occupying approximately the same position now held by the much better known Cretaceous Idaho Batholith (Fig. 8-4). Because large exposures of the augen gneiss occur on both sides of the Cretaceous batholith, it is possible that this Idaho batholith penetrated upwards along a zone of crustal weakness similar to that followed by the Precambrian batholith.

The augen gneiss bodies were intruded synkinematically into biotite gneiss and amphibolite. All contacts of the augen gneiss with the amphibolite and the biotite gneiss are sharp and conformable. Included within the augen gneiss bodies are numerous septa and xenoliths of the metasedimentary rocks. These augen gneisses are primarily composed of microcline, plagioclase, biotite and quartz.

Cataclastic texture is well developed with mortar and granulated zones especially at the edges of the microcline porphyroblasts. Biotite-rich zones curve around porphyroblasts and provide a strongly foliated appearance to the rock. Microcline porphyroblasts are lenticular and strongly flattened in the plane of foliation. Extreme flattening occurs locally and results in flaser texture (Fig. 8-4) with crystals less than one-half inch thick. Foliation is well developed with compositional layering defined by alternating feldspar-rich and biotite-rich layers.

Igneous Origin of the Matrix. The field evidence supports an igneous origin for the matrix. Contacts between the augen gneiss and the rocks it cuts are very sharp. There is not a gradual chemical or textural change but an abrupt transition from one rock type to another. Rotated xenoliths of amphibolite and biotite gneiss representative of the host rock are common in the augen gneiss. The zircon crystals are simple and sharply faceted.

Origin of the Microcline Megacrysts. Euhedral potassium crystals are a common phenomenon in gneisses even though potassium feldspar is at the bottom of the idioblastic series. In many cases potassium feldspar crystals, similar in appearance to those of the augen gneiss, have grown in the wall rock and in xenoliths of amphibolite and metasediments within 10 cm of the contact. These crystals in the wall rock are commonly the same size and shape as those in the adjacent augen gneiss body.

Rapakivi texture or the mantling of potassium feldspar megacrysts by oligoclase is common in the augen gneiss (Fig. 8-7). These mantles consist of myrmekitic plagioclase and range from a fraction of 1 mm to 4 mm thick.

Micaceous quartzite of Middle Mountain

In south-central Idaho a remarkable quartzite is mined from a group of quarries situated on the west flank of Middle Mountain, Idaho (Figs. 8-8A and 8-8B). This rock is unusual because it splits along a well-developed cleavage into large plates slightly more than 1 cm in thickness and 3 m in diameter (Figs. 8-9 and 8-10).

On the west side of Middle Mountain, a large tabular-shaped plate of white to dark gray micaceous quartzite is exposed. This quartzite layer ranges from 10 m to 100 m thick and is underlain by a granitic intrusion. The quartzite, which is more

306

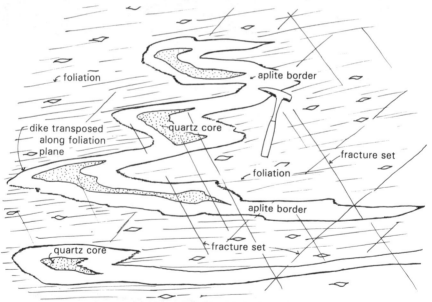

Fig. 8-5. Pegmatite dikes have appearance of offset by transposition along the foliation plane by ductile shearing. Note the well developed zoning represented by quartz cores and feldspar borders. Near Shoup, Idaho.

307

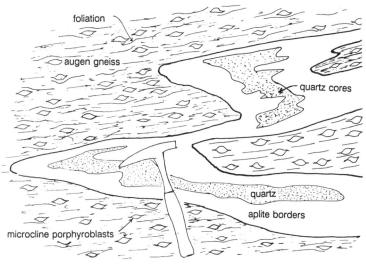

foliation

augen gneiss

quartz cores

quartz

aplite borders

microcline porphyroblasts

Fig. 8-6. Closeup of Fig. 8-5.

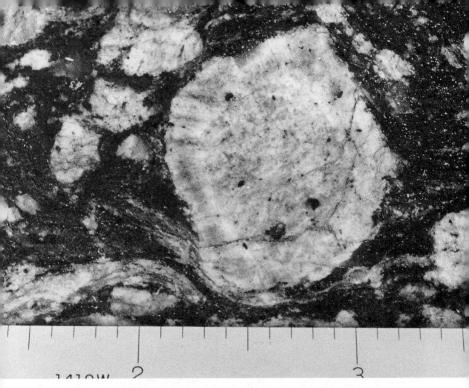

Fig. 8-7. Oligoclase mantle around microcline porphyroblast in strongly foliated augen gneiss. This is called **rapikivi** texture. Near Shoup, Idaho.

resistant than the underlying rock, tends to form a rim where it makes contact with the softer granite. The color of the quartzite ranges from pure white to dark gray. The clay layers in the original sediments were converted to coarse-grained micas, primarily muscovite and biotite. Compositional layering is exceptionally well developed in the quartzite. Thin, dark, mica-rich layers alternate with thicker, light, quartz-rich layers. The color contrasts between the thin layers of dark minerals and the white quartzite make it easy to identify and interpret mesoscopic primary and secondary structures. Primary structures such as graded bedding, compositional layering and cross bedding are common.

The following features indicate that the intrusion of the granite was mesozonal and synkinematic with the deformation: (1) abundant lit-par-lit injection of the granitic material and axial-plane dikes near the contact; (2) folding during and after the intrusion; (3) concordant contacts between the granite and quartzite (gneissic border zones in the granite); and (4) folded secondary foliation.

Fig. 8-9. An excellent example of spaced cleavage in micaceous quartzite. This cleavage is consistently spaced at 1 to 2 cm throughout the exposure. The cleavage is defined by layers of large muscovite grains up to 1 cm in diameter and is exceptionally well developed where parallel to bedding. Middle Mountain, Idaho.

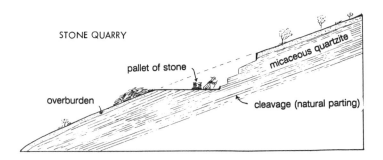

Fig. 8-8A. Cross section of quarry where micaceous quartzite is mined for use as a veneer building stone. Note that the topographic surface is parallel to the dip of the cleavage in the quartzite which gives a very significant advantage in extracting large-diameter plates from the quarry.

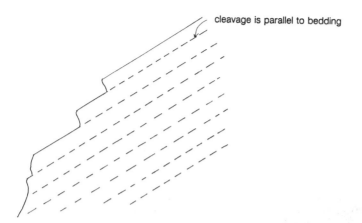

cleavage is parallel to bedding

Fig. 8-8B. Spaced cleavage in Middle Mountain Quartzite.

Fig. 8-10. Large, flat plates of quartzite up to 3 m in diameter and approximately 1.5 cm thick ready for shipment; plates are stacked vertically to minimize breaking. Middle Mountain, Idaho.

INDEX